The Attention Deficit

"This is a timely book on a pressing topic. Individually and collectively we face problems caused by the unintended consequences of digital connectivity, and Bhatt's book brings helpful clarity to these issues."
—Matthew J. Salganik, *Professor of Sociology, Princeton University*

Swati Bhatt

The Attention Deficit

Unintended Consequences of Digital Connectivity

Swati Bhatt
Department of Economics
Princeton University
Princeton, NJ, USA

ISBN 978-3-030-21847-8   ISBN 978-3-030-21848-5   (eBook)
https://doi.org/10.1007/978-3-030-21848-5

Cover illustration: © Ints Vikmanis / Alamy Stock Photo

This Palgrave Macmillan imprint is published by the registered company Springer Nature Switzerland AG
The registered company address is: Gewerbestrasse 11, 6330 Cham, Switzerland

To Ishaan

Preface

And thus the native hue of resolution
Is sicklied o'er with the pale cast of thought,
And enterprises of great pith and moment
With this regard their currents turn awry,
And lose the name of action.
—William Shakespeare, Hamlet

I write this book to explain why things are the way they are, why we live the lives we do and why we make certain choices in the face of rapid technological change. Why has the rhetoric about the rise of social media, privacy and tech behemoths escalated? Why is fear, mistrust and risk aversion on the rise? Why is entrepreneurship losing "the name of action," as in the quote above?

Attention is a scarce resource and emergent demands for this resource create a deficit, much like a budget deficit. The unique approach to attention in my book is in defining this resource in terms of time. Balancing the hours available for attention against the hours demanded by a tsunami of content creates a time deficit, called the attention deficit. Humans have never before been so profoundly networked and exchanged such vast quantities of information. Ubiquitous connectivity and sharing have unleashed a torrent of information and created an extravagant demand for mental effort. The very process of filtering information is effortful, placing additional stress on cognitive resources. At the same time, devices and algorithms map our preferences, generate predictions to optimize our lives, anaesthetizing and shrinking our available attention resources. Greater demand matched against a reduced supply of attention results in an attention deficit, which manifests as cognitive apathy or mental paralysis.

In the face of rapid technological change, a deficit of mental resources precludes adaptation. Adjusting to technological change is an effortful process since it requires reframing and rebalancing of lifestyles and world-views. In order to conserve scarce attention resources, there is resistance to change and a refusal to adapt in the event of change. This manifests as risk aversion. The principle thesis of this book is that fear, mistrust and risk aversion, driven by the attention deficit and cognitive apathy, have diminished creativity, entrepreneurship and risk-taking.

This book has grown out of courses on technology and markets that I have taught at Princeton over the past five years, and discussions with students both in class and outside. The intellectual energy brought by these young folks has invigorated and shaped my thinking. Their probing questions helped clarify my ideas and sharpen my message. I am deeply grateful to all.

In particular, David Kim, Caroline Lippman, Reed Malcionda, Elizabeth Petrov, Phoebe Rogers, Noah Schochet and Samantha Shapiro offered valuable feedback on early drafts, along with cherished conversa-tions, inducing me to write lucidly yet simply.

Discussions with Bevin Benson, Brandon Callegari, Zachary Dinch, Trevor Forbes, Alex Ford, Abigail Gupta, Bryce Mbanefo, Pooja Parmar, Jamie Rosen, Elias Stern, Ayushi Sinha, Ryan Yao, Yan Zhang, Cameron Zeluck and Katie Zhou were not only beneficial but injected a touch of much appreciated levity. Fernanda Macias, as a graduate student assis-tant, was incredible in facilitating class discussions.

Term papers for classes revealed deep insights by the students. Joseph Flynn, for instance, wrote about the unforeseen consequences of Facebook's Free Basics program, which gave users in Africa and South-East Asia free access to services but also incited violence against the Rohingya Muslims in Myanmar and uprisings in the Amhara and Oromia regions in Ethiopia; John Colangelo talked about podcasts, adding to the content tsunami by enabling multitasking; Claire Collins investigated how consumer preferences change when popular brands are associated with influencers on social media, pointing out that content derives from the person and not the brand; and Zachary Kuehm showed how technology is industrializing the production of music when the same set of people write songs for multiple artists.

The economics department at Princeton University has been my incredible home for the past 27 years, providing intellectual challenges in a secure open environment. My deepest gratitude is to Avinash Dixit, who taught me how to apply economic reasoning intelligently and wisely. He was my Ph.D. adviser, is now my lifelong friend and holds my highest regard.

Elizabeth Graber, my editor at Palgrave Macmillan, believed in my message and supported me with great equanimity throughout the process. Also at Palgrave Macmillan, Sophia Siegler's gentle reminders about the devil being in the details helped me polish my message. Barbara Radvany, Laura Sciarotta and Laura Hedden did graceful handholding at crucial and unexpected moments. Matthew Parker made the computer and networking logistics so much smoother.

Writing a book places challenges on one's family and mine was no different. But daily morning runs with my son, Ishaan, lightened the pace. As did planning and participating in my daughter Anjali's wedding as this book was going to press—we both agreed that the *best* need not be the enemy of the *good*. Ravin, my husband, developed the mental fortitude to see movies and attend music concerts unaccompanied, in solo mode, a process laced with much humor.

Princeton, NJ, USA   Swati Bhatt

Contents

List of Figures

Key Words

Attention = Cognitive bandwidth hours = hours of mental effort, analogous to the range of frequencies available for data transmission on the Internet

Free will = Agency in initiating and managing decisions to achieve a desired outcome, measured as supply of attention or cognitive bandwidth hours. Usually fixed at normal waking hours or 16

Individual liberty = Freedom to execute actions to achieve desired outcomes

Attention deficit = excess demand for cognitive bandwidth hours relative to available hours

Digital content = information transmitted online

Internet = (shorthand for) Architecture and protocols of mobile and fixed digital information and communications technologies (ICT)

Information and communication markets = exchange of digital content for payment

Introduction

Information and communication technology has enabled connectivity on an unimagined scale. Acknowledging humans as social animals, economic activity promotes this socialization. Market transactions are based on optimism and trust, as individuals invest in the future by having children, by extending credit and accepting risk, and by building connections in the sincere expectation of this connectivity being reciprocated. However, in excess, connections compromise entrepreneurship and risk-taking.

Ubiquitous connectivity has four effects. The *first effect*, transparency in human interaction, is captured by the sharing model. Digital representation of information offers multiple avenues for sharing content. Human experience has been enhanced and enriched by the dense web of connections and the outpouring of shared digital content. Sharing personal information lights up the brain because there are benefits of group affirmation and inclusion when personality traits are shared; we derive self-esteem by comparisons with other groups and individuals.

However, this sharing and comparing leads to judgment. Negative judgments corrode self-esteem, leading to anxiety and depression. Paradoxically, active engagement with social connections creates a vacuum, a loss of self-esteem as connections inevitably lead to comparisons with other groups and individuals. Restoring esteem leads to even more connections, sharing and comparisons.

Therefore, the *second effect* is commercialization of sharing, leading to a tsunami of content. Loss of self-worth, driven by the first effect, encourages further connectivity and sharing, as buyers seek more comfort, more reassurance, via social media, paying with time and personal information. The process of connecting and sharing, of exchanging content digitally, has become a market by commodifying each additional connection in terms of time.

Friendships are valued in follower counts, with each count implying that a friend has devoted time to your social media page. A larger number of followers connote a greater amount of time spent on your Facebook page. When time and content are decoupled in asynchronous communication, as when response time or delay on messaging apps is carefully controlled by users, multiple connections can be managed simultaneously. High follower counts, suggesting reliability, translate into personal wealth as individuals become paid influencers on Twitter.

The outcome of this content tsunami is an attention deficit, the *third effect*. Consumption of content utilizes time and attention, such that the product is digital content and the payment is with time and data. Correspondingly, social media fulfill this demand for content with exuberance, both via user-generated content and via commercially curated content. Not only does processing this vast quantity of information utilize available hours of attention but the need to filter this exorbitant content further captures the mind, exacerbating the scarcity of attention.

In addition, when devices and algorithms map our preferences and make predictions guiding decision-making, mental effort becomes superfluous. Not needing to exercise the mind, available hours of attention shrink.

The confluence of increased demand for digital content and reduced supply of cognition due to algorithmic prediction capabilities results in an attention deficit. This scarcity manifests as cognitive apathy, a mental paralysis equivalent to systemic failure of computer networks.

Cognitive apathy impairs judgment and decision-making, leading to the *fourth effect*, mistrust, fear and diminished risk-taking. Reorienting worldviews and lifestyles amidst the swirling winds of technology-driven disruption demands judicious adaptation. It takes some effort to incorporate new ways of doing things and absorbing new ways of perceiving the world. When mental resources are compromised, adjustment to change is resisted and risk avoidance is predominant as nostalgia and the familiarity of entrenched behavior take over. We observe declining entrepreneurship, innovation and imagination.

Recognizing human beings as social animals, the content tsunami powers two streams of consciousness—sharing and individual liberty. With the sharing frame of mind, there is transparency, trust, inclusion and cooperation. On the other hand, the individual liberty approach magnifies the personal factor, with its focus on the individual. The ensuing comparisons and judgments invoke notions of privacy. A conflict arises because risk-taking invokes trust, while privacy is modeled on risk fearing and mistrust. Anonymity, freedom to be and do, is needed *because* our imagination fears catastrophic outcomes. Despite the benefits of affirmation and group inclusion, connectivity elicits judgment, mistrust, fear, anxiety and depression. The digital revolution that brought us connectivity is exhibiting the unimagined consequences of tribal prejudice and isolation in echo chambers. Cognitive apathy follows, as there is no need to adapt to changing circumstances or different people. What started out as a voice for individual liberty can mutate into a loss of free will.

The past decade has seen a lengthening trail of technology criticism and warnings about the threats of commerce and machine persuasion compromising our free will. To be sure, the innovative medium that provides information can also be used to serve nefarious purposes. Inventions, including the technologies developed for splitting the atom, have multiple possible uses and one cannot blame the tool for threats issued by users of the tool. So also, digital communication technology is a tool for knowledge dissemination and we must partner with it as best as we can.

On the digital savannah, "we are all connected. …Everything you see exists together in a great delicate balance" as Mufasa says in *The Lion King*, Disney's 1994 epic. Preserving this balance of connectivity in a spirit of bold thinking is the way forward.

1

Connectivity, Attention and Risk

On a cold January weekend in 2019, at a conference at Princeton University, a recently matriculated physics undergraduate appeared fascinated by entrepreneurial possibilities. The conference was on emerging risks, opportunities and governance of artificial intelligence in environmental and agricultural applications. The undergraduate had developed an algorithm, in the emerging field of agritech, to optimize water use in drought-prone areas. His idea was to power small, lightweight drones with moisture-sensing ability for watering agricultural land. However, he had been derailed by job obligations and, importantly, what he called "life's distractions." What had held him back? Mental overload or mental laziness? Was mental overload due to a content tsunami, generated by ubiquitous connectivity, or was mental laziness engendered by loss of autonomy due to devices and software? How was this young undergraduate a beneficiary of digital information and communication technology; a technology that spawned deep connectivity, communication and machine-enabled prediction and thinking? Why was it difficult to translate technology into action?

Humans care about control and autonomy over their lives. This is the idea behind free will and the maximization of utility. Consequently, the business model is based on trading data for free and personalized products

© The Author(s) 2019
S. Bhatt, *The Attention Deficit*, https://doi.org/10.1007/978-3-030-21848-5_1

and services. But individuals also care about acknowledgment and legitimacy in who they are and what they do. Therefore, when self-revelation provides esteem and a sense of personal identity, individuals voluntarily share personal data.

This sharing model leads to a content tsunami. The capacity to comprehend is overwhelmed by this exorbitant demand for cognitive faculties and the outcome is the attention deficit. Human faculties were physically imprinted at least 50,000 years ago in the *Big Bang of Human Consciousness*, so our ability to attend to information is a scarce resource.[1] The process of applying filters and sorting through this tsunami further strains cognitive capacity. A resource deficit arises when the demand for attention exceeds this scarce supply.

Furthermore, artificial intelligence incorporated in devices automates day-to-day decisions by algorithms that organize information and make predictions and recommendations. Ceding agency to code shrinks available mental effort and weakens the capacity to adequately filter information, promoting mental atrophy. An overwhelming demand for attention combined with a weakened filtering capacity and scarce cognitive bandwidth has spawned a mental framework of *cognitive apathy* that does not support a wider vision of responsibility and risk-taking.[2]

Fear, mistrust and risk aversion are pervasive. Technologies that threaten our worldview and lifestyles by requiring adaptation and adjustment are faced with resistance. Widespread apathy is the response, and in the context of an increase in demand for our attention, there is impaired judgment and loss of decision-making skills.

---

[1] After the particularly harsh ice ages, spanning 190,000–90,000 B.C.E., eastern and southern Africa became warmer and wetter, according to Ian Morris. "By 50,000 B.C.E. modern humans were thinking and acting on a whole different plane from their ancestors." The *Great Leap* around 50,000 B.C.E. "began with purely neurological changes that rewired the brain to make modern kinds of speech possible, which in turn drove a revolution in behavior" (Morris 2010). Neuroplasticity allows the mind to adapt to new environments, but the physical dimensions of the brain have remained unchanged.

[2] While I define cognitive bandwidth hours in terms of time, the unit-free umbrella term—bandwidth—was introduced by Mullainathan and Shafir in their book *Scarcity*. Bandwidth is a generic term for a scarce resource: computational capacity or mental capacity, and it encompasses "fluid intelligence, a key resource that affects how we process information and make decisions" as well as executive control or impulse control (Mullainathan and Shafir 2013).

The young man introduced earlier in this chapter was overwhelmed with distracting information. Having helped his friend launch a successful fitness application, he was anxious about keeping in touch with academic research that would impact his fledging idea and so he attended conferences, read voraciously and networked on all fronts. Amidst concern about his father's uneasy financial situation and anxious about geopolitical uncertainty, he had caved into a sense of fearfulness and insecurity about his own future. In other words, he had decided to "wait it out."

The reality is consistent with such anecdotes. About a third of college students reported feelings of overwhelming anxiety and over two-thirds felt overwhelmed by their responsibilities in 2018 (American College Health Association Survey 2018. There is a drag on economic dynamism as seen in a startup deficit and decline in seed funding; a rise in economic behemoths; cultural nostalgia and a reduction of civic awareness. Let me explain in terms of four forces, four facts and four aspects of the sharing model.

Four Forces

The first of the four forces unleashed by information and communications technology (ICT) is connectivity.[3] Connectivity between individuals allows sharing, the transfer of information between individuals seamlessly and at nearly zero cost. Connections are being made and reinforced across the human network on an unimagined scale and information is being shared with abandon.

Second, the resulting content tsunami has led to an exorbitant demand for attention, defined as cognitive bandwidth hours available for mental effort. Filtering the vast quantity of information imposes additional demands upon mental faculties. The tsunami of content, while contributing to an attention deficit, has another perverse outcome.

---

[3] According to my colleague at Princeton, Brian Kernighan, who contributed to the development of Unix and multiple programming languages while at Bell Labs, digital information and communications technology encompasses universal digital representation of information plus universal digital processors (computers) plus universal digital networks and massive amounts of digital data (Kernighan 2018). Artificial intelligence is a general-purpose technology and an input in the production of ideas and goods. Machine Learning (ML), a subset of Artificial Intelligence (AI), addresses prediction, based on historical or experimental training data (Agrawal et al. 2018).

Sharing information opens the door to comparisons and judgment, where fear, anxiety and mistrust dominate. Retreating behind the rhetoric of privacy, the second unanticipated consequence is tribal thinking. We observe prejudice, narcissism, isolation and solitary consumption. Such an environment is inimical to risk-taking. Paradoxically, sharing itself connotes a transparent, trusting landscape where bold ideas can be confidently explored.

Artificial intelligence, which is connectivity between humans and machines, has liberated us from the inconvenience and decision-making involved with daily, mundane tasks. Recommendation algorithms simplify decision-making. When, for example, Waze maps out a route from point A to point B, we are relieved from the mental effort of both remembering and optimizing details of the route. Fitness apps have trainer-led programs that combine workouts with close monitoring of our sleep/wake cycles, our nutrition and diet schedules and our calendar of activities so we no longer have to figure out how to structure our lives. We don't even have to leave the familiar couch since Alexa, the device from Amazon, can decipher our tastes and implement our favorite entertainment. Paradoxically, the larger the content tsunami, the greater the reliance on algorithms to assist in processing this information. Effectively, this respite has shut down our mental faculties, shrinking the supply of cognitive bandwidth hours.

Third, juxtaposing the increase in demand for available cognitive hours with the reduced supply of hours, there is a net deficit of cognitive bandwidth hours or an attention deficit. Figure 1.1 illustrates individual desire for cognitive bandwidth increasing as a function of the content tsunami, and capacity or available bandwidth decreasing as content tsunami increases. The diminishing capacity is due to mental redundancy created by algorithmic outsourcing. When the tsunami is large, there arises an attention deficit as shown by the vertical blue line in the diagram. Demand for bandwidth exceeds the supply, and this deficit is magnified by increasing the tsunami from level 3 to level 5 or higher. Both desire and capacity for available cognitive bandwidth are monitored to a varying degree by individuals (see Chap. 6 for details). Define free will as agency in initiating and managing decisions to achieve desired outcomes and individual liberty as the freedom to execute these decisions. Then, the attention deficit suggests a loss of free will or consumer sovereignty as cognitive faculties are compromised, and we become incapable of simple decision-making.

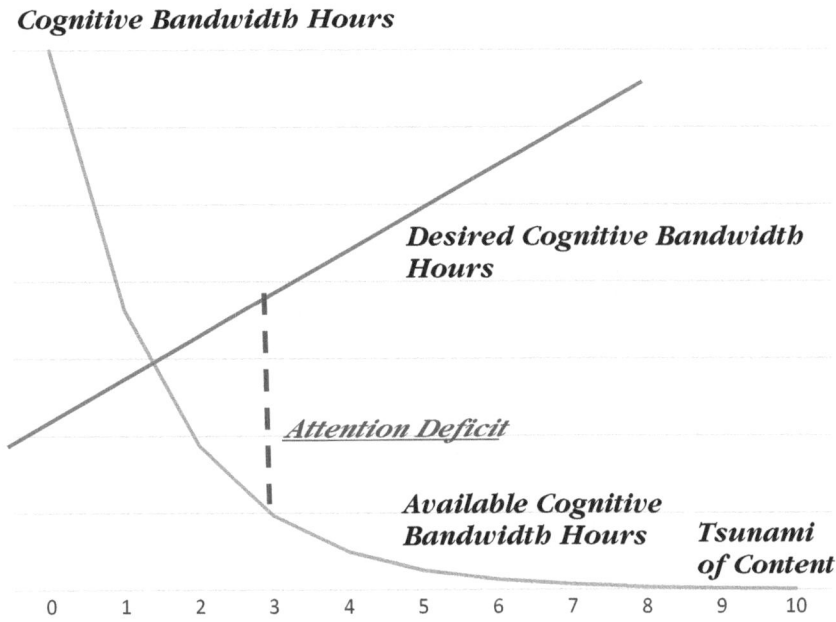

Fig. 1.1 Attention deficit = deficit of cognitive bandwidth hours. (See Chap. 6 for details. Source: Author's calculations)

Fourth, in the face of change, adaptation requires mental effort, whose scarcity leads to inaction. This is cognitive apathy. Adapting to a changing environment generated by new technologies demands resourcefulness and resilience. In the face of uncertainty, decision-making involves a creative assessment of outcomes and their probabilities, a specification of how resources can be matched to produce favorable payoffs, and an imagination capable of out-of-the-box thinking. Mental effort is required to assess new situations and when available cognitive capacity is compromised a budget-tightening mindset evolves.

Resistance to change and a reluctance to adapt is a way of conserving cognitive bandwidth. This is risk avoidance. Declining entrepreneurship, creativity and imaginative business ventures are economic consequences of resistance to change, while nostalgia and tribalism are the social outcomes. Since adaptation is a process requiring deliberation and cognitive processing, inadequate processing capacity compromises free will and manifests as cognitive apathy and risk avoidance. The central theme of this book is that the confluence of a content tsunami, artificial intelligence and attention deficit has the unintended consequence of diminished risk-taking.

Four Facts

Now consider the four facts. First, we are in the throes of climate uncertainty on a global scale. Long-term temperature response to greenhouse gas (carbon) emissions, called the carbon-climate response (CCR), depends upon the carbon sensitivity, the increase in atmospheric CO_2 concentration from carbon emissions and climate sensitivity, which is the warming generated by a certain amount of carbon emissions (Brock and Hansen 2018).

Second, we are witnessing a rapid escalation of income and wealth inequality. Regardless of our attitudes toward equal opportunity, equal outcomes with regard to income distribution, level playing fields and fair rules of the game, we still need to have a dialogue about where we are headed. Inequality and climate uncertainty are intertwined in their global ramifications, suggesting an urgency in geopolitical discussions.

Third, ICT has created new patterns in our lives with new markets, new products and new methods of payment. One new market is the market for friendship, where the metric is follower counts, retweets, likes and shares. We bid for high follower counts by Instagramming, YouTubing, Facebooking and tweeting personal information, in exchange for free content, both traditional and branded. Without the constraint of physical, embodied interaction, we communicate in isolation and asynchronously: texts and replies are not simultaneous and interactive since individuals determine the response time lag. Free content turns into a content tsunami, striking against the wall of scarce cognitive bandwidth.

Fourth, there is fear, mistrust and risk aversion. There is mistrust of people and institutions and a desire for the safe haven of nostalgia and familiarity of past cultural landmarks. To illustrate fear, take the case of parenting styles. Child-rearing today is *hothousing* children with structured play, emphasis on rote learning and detailed time management of children's life. For example, self-reported quarterly spending per child under the age of six in the highest income decile in the US increased from $3000 per annum in the 1970s to $9000 per annum in 2010, with little change in the lower deciles, based on before-tax income (Kornrich 2016).[4] This raises the possibility that parents are unwilling to risk their child's future financial outcome to innate skill and a more relaxed parenting style. When "globalization has cranked up competition for the best jobs," even an excursion by nursery school children to a "forest experience center" is considered risky and "some parents stop their offspring from taking part in these excursions for fear that they might get hurt" (The Art and Science of Parenting 2019).

Admittedly these children are growing up in the post-9/11 era and have been exposed to traumatic economic changes (the recession of 2008) and political uncertainties. According to a 2018 survey of 19,664 college students by the American College Health Association, anxiety is the most common student mental health problem today, with 62.9 percent of all students reporting incidence of anxiety within the previous year and 27.4 percent of students reported that anxiety had affected their academic performance in the previous 12 months, while 42.7 percent reported having felt so depressed that they had trouble functioning in the previous 12 months. However, when 86.4 percent of college students report feeling overwhelmed by all their responsibilities in the past year, one suspects that a content tsunami, an information overload, might be the underlying cause.

Risk aversion is pervasive as seen in the decline in startups and entrepreneurship activity (as detailed in Chap. 7). Growth in intellectual property products, an investment category in national income accounts, as a fraction of gross private domestic investment has been relatively flat since 2014 (see Fig. 1.2). Anecdotal evidence points to a changing mindset among startup founders, who are moving away from the high-pressure world of established venture capital, trying to find funding at smaller scales and from nontraditional sources. Some founders, unable to meet the stringent criteria set by investors, "wither[ed] under the pressure of

---

[4] The study used 1972–1973 and 1980–2010 inflation-adjusted data from the Consumer Expenditure Survey conducted by the Bureau of Labor Statistics.

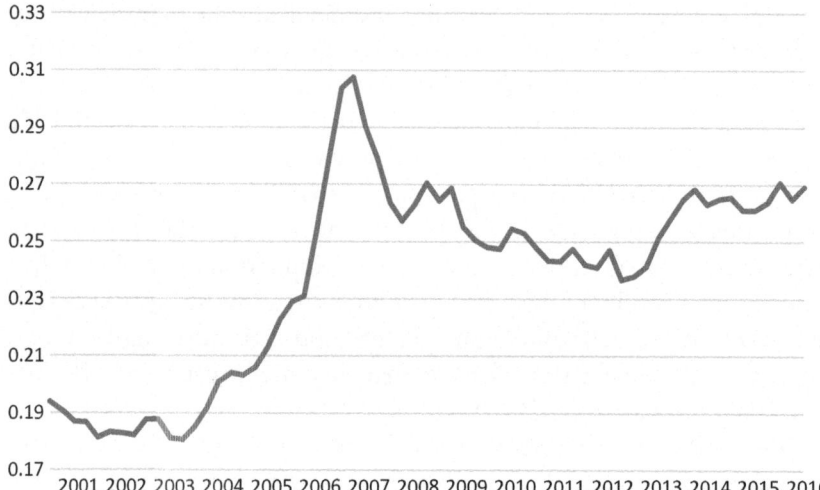

Fig. 1.2 Intellectual property products over gross private domestic investment, seasonally adjusted, chained 2012 dollars. (Note: Intellectual property products is a line item under gross private domestic investment, which is the sum of nonresidential investment, residential investment and change in private inventories. Gross domestic product is the sum of personal consumption expenditures, gross private domestic investment, net exports of goods and services, government consumption expenditures and government gross investment. Source: Data from the Bureau of Economic Analysis, US Department of Commerce. National Data: GDP and Personal Income, Table 1.1.6. Accessed 4/15/19 from https://apps.bea.gov/iTable/iTable.cfm?reqid=19&step=2#reqid=19&step=2&isuri=1&1921=survey)

hypergrowth" (American College Health Association National Survey of 19,664 college students n.d.).

Evidence of risk aversion is the decline in business dynamism, a startup deficit and a stagnant labor demand since the 1990s. If ICT is to make a difference, it should be possible to produce more goods and services with the same amount of labor, what is called *labor productivity*, and jobs should be reallocated from low-productivity sectors to those with higher productivity, a process called *business dynamism* and a growth in labor demand. But "something changed around 2000." Even star firms—the top 20 firms by market value—are not contributing to overall productivity growth compared with the star firms from two decades ago (Griffith 2019).

Productivity and Dynamism Data from the Bureau of Labor Statistics show that the average annual growth rate of labor productivity (measured as output per hour worked) in the private non-farm business sector (which accounts for about 90 percent of hours worked) was 1.4 during 1987–1990, rose to 1.7 from 1990 to 1995, further increasing to 2.7 from 1995 to 2007 and then dropped to 1.2 during 2007–2017. This "is consistent with the notion that post-2000 declining business dynamism has not been benign for American living standards, but, instead, is closely related to slowing productivity growth" (Decker et al. 2018). Moreover, the evidence suggests that new business creation and responsiveness to differences in labor productivity both within industries and across industries have shown pervasive declines since 2000. Decker et al. show that "business growth and survival has become less responsive to idiosyncratic productivity (both TFP and labor productivity) with an especially large decline in responsiveness for young firms in the high-tech sector" (Decker et al. 2018). Between 1999 and 2016, the employment to population ratio for each 10-year age group in the 25–54 age range dropped 3–4 percent. The drop was 5.6 percent for men aged 25–34 (Decker et al. 2018). In other words, there has been a sluggish response of firms in their employment and hiring decisions to disparities in labor productivity across their industry. Gutierrez et al. show that superstar firms, defined as the top 20 firms by market value in any given year, "have not become more productive" and that "it is clear that something changed around 2000" (Gutierrez et al. 2019).

Startup Deficit The aging of the US business sector has increased due to declining startup rates. Alon et al. have shown that there is a persistent and widespread shrinkage in new firm entry, which they define as the startup deficit. In particular, "the start-up deficit and the subsequent aging of the US business sector have had a considerable impact on aggregate productivity," reducing aggregate productivity by "roughly 0.10 percentage points a year from 1980–2014. While the per annum rate is small, the cumulative effect over the whole period is substantial, reducing the level of aggregate productivity by 3.1% by 2014" (Alon et al. 2018). The fall in startup rates negatively impacts productivity growth since there are fewer young high-productivity firms replacing mature, low-productivity firms. Reallocation of workers to the young high-productivity firms is then adversely affected.

A recent study shows that debtor protection laws in the US encourage banks to reduce credit to young firms in response to the expectation that marginal entrepreneurs, those needing credit, are more likely to fail. This is a self-fulfilling expectation in that lack of financial support makes exit inevitable. As a result, potential startups are deterred in their entry and venture funding decisions (Alon et al. 2018).

Stagnant Labor Market The employment share of young firms has dropped. Not only are there fewer startups but they employ fewer people on average. Shambaugh et al. find that employment shares by both firm age and firm size have declined. In 1987 both young firms (ages 0–10) and small firms (less than 50 employees) constituted 33 percent of employed labor force. By 2014 this share had dropped to 19 percent for young firms and even further to 27.2 percent for small firms (Shambaugh et al. 2018).

Over the past 30 years, labor demand has shown a secular decline. "Labor demand grew on average by 2.4% per annum between 1947 and 1987, but then slowed down to a growth rate of 1.33% per annum from 1987 onwards, and has been essentially stagnant since the late 1990s." Acemoglu and Restrepo find that the slowdown in growth of labor demand is due to weaker productivity effects and a *decrease* in the task content of production that would favor labor. A large proportion of technological change over the past three decades, and especially since 1990, came from accelerated automation, which meant the replacement of capital for labor for certain tasks. However, this was not offset by "reinstatement effects coming from new technologies" or the creation of new tasks, as measured by new occupations or the emergence of previously nonexistent job titles (Acemoglu and Restrepo 2018).

On the other hand, perhaps there is some latency in outcomes as we await the development of complementary and upper floor technologies which might unleash the power of mezzanine technologies. For example, the introduction of the iPhone in 2007 did not immediately increase the value or outside impact of the device. The proliferation of applications occurred after Apple opened up the iPhone operating system (iOS) platform to developers in 2012 and the potential of the device was gauged. You could conduct business with just the single smartphone, replacing the desktop. Streaming technology is a case in point. While originally developed in the early 1990s, Netflix matched streaming technology with the mobile dimension in 2012 to unleash its content on users. In less than a decade, Netflix has become one of the ten largest companies in the US, in terms of market capitalization.

A number of hypotheses have been put forward to explain this decline in startup activity, responsiveness and job reallocation such as the internal dynamics of innovation, the rise of behemoths and general market frictions. A period of rapid innovation leads to new firms and productivity growth, followed by a slowdown as implementation of these technologies slowly takes effect (Acemoglu and Restrepo 2018). Network effects and economies of scale in distribution and supply chains have propelled behemoths, which are purported to thwart emerging businesses simply by integrating them into the fold. Finally, regulatory and financial frictions make markets less hospitable, but technological advances in fintech and current developments in the regulatory landscape appear to have mitigated these drawbacks. Occupational licensing imposes entry restrictions into many occupations as well as reducing geographic mobility. Relicensing after an interstate move acts as an additional impediment to worker relocation in response to the high-wage job opportunities (Cerqueiro et al. 2018).

The parallel existence of granular peer-to-peer firms and organizational behemoths is threatened with the rise of firms like Amazon, Google and Facebook. Leveraging network effects and the internal dynamics of production technology (i.e. increasing returns to scale, and AI-enabled data analytics), these behemoths have captured ever-increasing shares of their respective markets.[5] Traditional anti-trust arguments, that prices-cost margins are too large, are not entirely valid today. Success of these behemoths, as measured by sales revenue, suggests that these firms are efficient

---

[5] Economies of scale, and declining average costs, will then arise if input costs are constant as the firm expands scale.

at seeking out, aggregating and reflecting popular culture. Importantly, these behemoths, aided by low interest rates, are more likely to swallow nascent startups as they look to create a ubiquitous ecosystem. Netflix establishes its hegemony by leveraging the power of network effects and algorithms. It notes viewer preferences by tabulating the frequency and number of movies watched, the completion rate and other descriptive numbers, which are then processed by complex algorithms across the universe of users in order to make recommendations.

Summarizing, ubiquitous connectivity and sharing have created an extravagant demand for mental effort. Simultaneously, mental atrophy promoted by devices and algorithms has shrunk the supply, so the result is an attention deficit, which manifests as cognitive apathy. *The principle thesis of this book is that mistrust, fear and risk aversion, driven by cognitive apathy, have diminished creativity, entrepreneurship and risk-taking.*[6]

There is one issue that remains unaddressed. A question that logically precedes the four dynamics is: why do we share in the first place? In the following sharing model, I describe how social transactions fuel both demand and supply in the market for content.

Four Aspects of the Sharing Model

The discussion is framed around four questions about the individual demand and supply of attention: Why do we share? How has time become the defining framework for sharing? What is the outcome of the content tsunami? How do content filters influence assessment of uncertainty?

(i) Why do we share? Human experience has been enhanced and enriched by the dense web of connections and the outpouring of shared digital content. In a pioneering study, Tamir and Mitchell (2012) found that "self-disclosure will act as an intrinsic reward" and that "self-disclosure

---

[6] The opposite of *cognitive apathy* is *buoyant market sentiment*, a term used by macroeconomists to describe markets where credit returns are below historical norms. Market sentiment reflecting risk appetite or beliefs about default probabilities in credit markets is used to identify credit expansion in the business cycle, since credit booms are thought to be a premonition of poor future economic performance (Gordon 2016).

was strongly associated with increased activation in brain regions that form the mesolimbic dopamine system." This is the same region of the brain that responds to food, money and social rewards such as humor since dopamine is the neurotransmitter responsible for the general feeling of well-being. In another paper, based on a series of experiments, Tamir and Mitchell found that participants "showed a significant preference for answering questions about self" over other alternatives and "were willing to forego money to think and talk about themselves" (Tamir and Mitchell 2012). Individuals self-disclose for two independent reasons: talking about oneself and introspection, which is self-referential thought. The authors conclude that humans not only like to think about themselves, they also like to talk about themselves.[7]

Sharing personal information lights up the brain because there are benefits of group affirmation and inclusion when personality traits are shared; we derive self-esteem by comparisons with other groups and individuals. Sharing reinforces the notion that one's ideas and experiences "can serve as a model for the experiences of others" allowing one to infer others' feelings (Tamir and Mitchell 2013). However, this sharing and comparing leads to judgment. Negative judgments corrode self-esteem, leading to anxiety and depression. Person A could experience negative feelings, and loss of esteem, upon viewing B's Instagram page and carefree vacation pictures. There follows a spiral, leading from connectivity to information exchange, which leads to comparisons and judgments. Ironically, this loss in esteem is salvaged by seeking more, newer connections, creating a flywheel of sharing and connecting.

To be clear, sharing can be a broad source of information, addressing diverse ideas and promoting thoughtful deliberation. In an open society it can encourage civil discussion that strengthens our ethical foundations, as is suggested by Alexander Hamilton in Federalist (Hamilton n.d.):

---

[7] The mesolimbic dopamine system includes the nucleus accumbens (NAcc) and the ventral tegmental area (VTA). In research studies, these brain regions showed activity under functional magnetic resonance imaging (fMRI) scanning when individuals engaged in self-disclosure, as well as when individuals were "merely introspecting about self" (López-Salido et al. 2017).

The differences of opinion, and the jarrings of parties in that department of the government, though they may sometimes obstruct salutary plans, yet often promote deliberation and circumspection, and serve to check excesses in the majority.

Content can enhance well-being and economic productivity, but it can also simply be a meandering river of content that transfers debris from one end to the other. Today, many Americans receive their news from social media—public perceptions are guided by political drama and fake news. As Alan Blinder puts it in his recent book:

The case of staged media events is particularly perplexing, since it is far from clear who is using whom. The top echelons of government obviously see such occasions as opportunities to sell their story to the public, using the news media as the indispensable intermediaries. But these same events invite the media to use the drawing power of leading politicians as stage props to help sell newspapers or boost TV ratings. (Blinder 2018)

And then there is the additional problem of fake news. Vousoughi et al. define "news as any story or claim with an assertion in it and a rumor as the social phenomena of a news story or claim spreading or diffusing through the Twitter network." Posts on social media are therefore more likely to be rumors than news, and filter bubbles, information that conforms to a users' prior beliefs, may start out as news but are then supported by rumors. A rumor cascade on Twitter is then a chain of retweets. These researchers found falsehoods were retweeted 70 percent more often than the truth. However, those who spread rumors were less socially active compared with those who spread news. They had significantly fewer followers, followed fewer people and had been active on Twitter for a shorter time period. "Falsehood diffused farther and faster than the truth despite these differences not because of them" (Vosoughi et al. 2018). False news tended to be novel and inspired disgust more often than sadness or trust, so the dispersal of rumors is driven more by human pathology than by underlying network structure.

(ii) How has time become the defining framework for sharing? In the aggregate market for communication and content, the commodity traded is follower count, or more generally digital content in the form of sharing, and the price is time. The process of connecting and sharing, of exchanging content digitally, has become a market by commodifying each additional connection. Familiarity and strength of connections is measured by time spent perusing their content. Thus, payment is in terms of time and communication markets facilitate exchange of digital content.

Connecting with people engages our time and attention, so standardizing engagement is the centerpiece of communication markets. The commodity is created by assigning values to "friends" with metrics such as numbers of Facebook followers (and likes), Instagram followers, Twitter followers (and retweets) and the size of subreddit communities, which make up the backbone of Reddit.com, the social news site. A larger number of followers connote a greater amount of time spent on that Facebook page.

As a twist on multitasking, there is asynchrony in the way we engage with the world, when, for example, response time or delay on iMessage or WhatsApp is carefully controlled by users. When time and content are decoupled in asynchronous communication, multiple connections can be simultaneously maintained. Frequency of interactions is measured by streaks on Snap, where a streak number is the days of unbroken snaps between two individuals. Snap also ranks friends by frequency of snaps and places a pink heart next to the top-ranked friend and an ugly emoji next to a non-reciprocated friendship listing. The ubiquitous use of Calendars coordinated across all of an individual's devices, and often their friends' devices, sets a new standard by defining one's life in terms of hours and further supports asynchrony. Every minute is accounted for in this calendar and pinging reminders are like the ka-ching of the cash register. Writers of pop songs have been preoccupied by the skip rate, the time it takes someone on Spotify to click to the next song. The mandate for songwriters is to get to the chorus in less than 30 seconds and make the introduction 2 seconds long (Vosoughi et al. 2018). Teenagers are best at asynchronous connection: they maintain multiple messaging channels while listening to music and doing their homework.

In fact, follower counts on social networks determine labor market outcomes, such as salary and job description. These numbers translate into personal wealth, as

> [t]he world's collective yearning for connection has not only reshaped the Fortune 500 and upended the advertising industry but also created a new status market: the number of people who follow, like or "friend" you. For some entertainers and entrepreneurs, this virtual status is a real-world currency. Follower counts on social networks help determine who will hire them, how much they are paid for bookings and endorsements, even how potential customers evaluate their businesses or products. (Leight 2017)

High follower counts, used as a metric for popularity, also suggest reliability and draw more followers. Influencers, for example, or individuals with a follower count of one million or more on Twitter earn $20,000 for a single promotional tweet (Leight 2017).

In the aggregate market for communication and content, as depicted in Fig. 1.3, the commodity traded is digital content in the form of sharing, and the price is time. Consequently, on the demand side, individuals buy content and pay with time and, *co-incidentally*, personal data. More precisely, when the time value of content falls buyers 'purchase' more content since cheaper content signifies ease of comprehension. when the time value of content falls buyers "purchase" more content since cheaper content signifies ease of comprehension. When buyers designed to induce greater time spent on that site, as shown by the supply curve in Fig. 1.3.

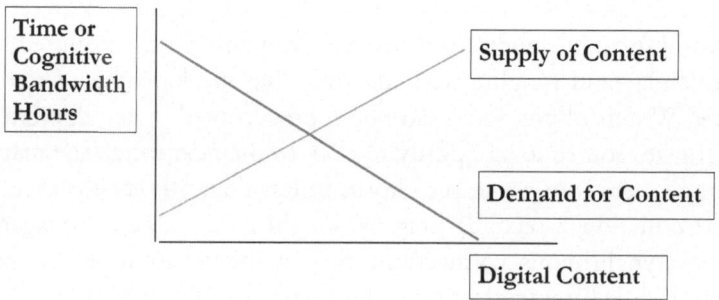

Fig. 1.3 Aggregate market for digital content. (Source: Author's image)

The more time a person spends browsing various online sites, mobile or fixed, more personally identifying information is revealed. Consequently, at the aggregate market level, the social media platform performs a "money laundering service" by selling you content in exchange for time, which is converted into data and sold for real currency to advertisers. Contrast Fig. 1.3 which depicts overall market demand and supply of content with Fig. 1.1 where demand and supply are from the perspective of the same individual: while one may want to consume more content (represented by the upward-sloping demand for cognitive bandwidth hours), one can only consume according to one's processing capacity and the limits of time (or supply of cognitive bandwidth hours).

User-friendly implementation of digital communication technology and a cognitively friendly format on social media platforms are the primary drivers of demand for content. When the feeling of *liking* communication becomes one of *craving* communication, addiction sets in as a driver of demand. Social media use has been shown to be correlated with extraversion and neuroticism; extraverted individuals are naturally social beings and drawn to social media. However, when a sense of urgency due to fear-of-missing-out (FOMO) is factored into the analysis, FOMO was a significant predictor of addiction.[8]

An obvious question to be asked at this point is how can individuals pay with cognitive bandwidth hours or attention when there are no clear property rights over personal time? What if this attention is appropriated without the owner's consent? The answer to this question raises important public policy issues, which are discussed in Chap. 9.

(iii) What is the outcome of this content tsunami? In the individual market for content, attention deficit results from overwhelming content as individuals are obliged to delegate time to internalize this content. With communication markets driving a content tsunami, consumption of or processing this information utilizes time. Importantly, mental effort is exercised in order to digest the vast amount of information, resulting in an attention deficit (see the toy model described

---

[8] Przybylski et al. write that FOMO, "defined as a pervasive apprehension that others might be having rewarding experiences from which one is absent, FOMO is characterized by the desire to stay continually connected with what others are doing" (2013).

in Chap. 6 for details). If pursuit of self-esteem and peer esteem drives most people to share text, and images in particular, who is the actual consumer of this content and what is the medium of consumption? Who reads all the text, who peruses the images and videos, and who internalizes and processes this content? Who has the time? Furthermore, the observed proliferation in content may be associated with an increased dispersion in quality, so content is presented as noisy information.

Effectively, therefore, social connections impinge upon a scarce resource—attention. In order to deal with the demands of the content tsunami on scarce cognitive capacity, we exercise selectivity in addressing information. The need to filter content generates further demands on this scarce resource. Writing about the psychology of scarcity, Mullainathan and Shafir emphasize that managing physical scarcity of any sort usurps attention or "captures the mind." Being focused on the scarce resource, "we have less mind to give to the rest of life." Exclusive focus can also be interpreted as tunnel vision, blinding us to any other issue (Mullainathan and Shafir 2013).

"Scarcity reduces all these components of bandwidth—it makes us less insightful, less forward-thinking, less controlled." With physical scarcity, "being poor, for example, reduces a person's cognitive capacity more than going one full night without sleep. It is not that the poor have less bandwidth as individuals. Rather, it is that the experience of poverty reduces anyone's bandwidth."

These filters are derived from the strict time constraint of 24-by-7 hours at our disposal and our underlying preferences. But herein lies a curious circularity. The filters provide us with information that we *want* to see, not information that we *should* see. Not only does this result in echo chambers, but it circumscribes the very environment that influences our preferences. We seek out the very information that seduces us to *want more*.

Managing this scarcity impairs the very judgment necessary to effectively separate the noise from true content. As we optimize content consumption against the constraint of attention, our effort telescopes over urgent content, displacing important content. Consequently, content consumed is narrow and urgent, not profound or important.

The idea of attention deficit has been widely researched and written about, noting that when slow, leisurely reading is replaced with faster access via digital representation of information, brain functioning could be altered (Przybylski et al. 2013). The literature on information and communications technology has focused on devices—their ubiquity and their digital format. The more basic question is how ICT-inspired connectivity has shaped our imagination. Asynchrony, a decoupling of time and content in an effort to create a content paradise, has created a content tsunami with time as the new metric. We therefore need a broad canvas, addressing the *phenomenon* of connectivity which encompasses not just *digitization* of information, but *access* to information and its consequences.[9]

(iv) How do content filters influence assessment of uncertainty? Objective assessment of uncertainty requires imagining all possible outcomes, postulating plausible relationships between scarce resources and outcomes and then assigning probabilities to these uncertain outcomes. This is a complex problem requiring judgment or appraisal, not just computational power. Scarce mental faculties are called upon to evaluate such situations so that heuristics are applied to conserve cognitive bandwidth. Absent even the minimal bandwidth for these mental shortcuts, the response to changing circumstances and situations of risk is apathy, an aversion to change accompanied by a reluctance to adapt.

Under conditions of risk and uncertainty, behavioral economics posits the frequent use of heuristics or the substitution of an easier question for a more complex one, to minimize cognitive effort. Not only is filtering required to manage scarce cognitive bandwidth, individuals are even

---

[9] The criticisms start with Nicholas Carr's *Is Google Making Us Stupid? What the Internet is Doing to our Brains* (2008), Tim Wu's *The Attention Merchants* (2016) and James Williams' *Stand Out of Our Light* (2018) (Carr 2008; Wu 2016; Williams 2018). These books take a sweeping view of the cultural zeitgeist, of the coherence of free will and democracy in a global media culture that mines psychological proclivities using the tools provided by ICT, which Williams calls industrialized persuasion and adversarial persuasion. These books suggest, from a philosophy of technology perspective, that there is an inadvertent design goal, which removes freedom of the mind, or cognitive autonomy in matters of importance. As discussed in the Appendix, free will as the source of democratic legitimacy is itself ambiguous—what is free will, is it context-free in matters of importance? (see the discussion in this chapter). Technology is neither our adversary nor does it have to provide a blueprint for navigating life. That is our job, as free human beings.

more likely to pay selective attention to salient information when making decisions and forming judgments when confronted with uncertainty (Carr 2008). Under normal conditions individuals prefer situations of cognitive ease. But when people are stressed under a content tsunami, these heuristics take on an added charge. We use judgment heuristics when making complex risk assessments. To simplify the problem, we arbitrarily constrain the available information set so that fewer steps are required for decision-making.

Under the *availability* heuristic, when more instances of the current question come to mind, frequency becomes the relevant decision criteria, replacing objective statistical measures. The importance of an event is often judged by the fluency and emotional charge with which similar events come to mind. I would fear downed trees during an upcoming hurricane, if there were recent reports of tree-related deaths, for example. An availability cascade is a self-sustaining chain of events, which may start from media reports of an event, lead up to panic and even public hysteria, with calls for government action. Content can influence ideas, and when stories are told via moving images on the screen, they have the power to rearrange mental frames of mind. A content tsunami covering a recent hurricane might trigger panic rather than a clear focus on the facts.

The associative mind traces out a coherent pattern of ideas from memory so the *associative* heuristic leads to stereotyping or representativeness. Facts that do not align with received wisdom will be ignored and outcomes may be distorted as confusion and anxiety increase risk perceptions and swing choices toward less risky options (Williams 2018).

Research in cognitive psychology suggests that attention accounts for a large proportion of the variation in decision-making even when all decision-makers are provided with identical and relevant information. In making decisions and forming judgments, a person is likely to pay selective attention to this information by creating content filters (Williams 2018). A universal desire for causality makes individuals susceptible to *framing* of the decision choices (order of options) and to *priming*, which forces an internal consistency to decisions.[10] A divide between the physi-

---

[10] "Decision makers face a wealth of potentially relevant information in the external environment and memory. Given the processing limitations of Homo Sapiens, selectivity is a central component

cal and online world is created due to perceived anonymity and desire for *confirmation bias*, so again information is selected so as to coincide with long-held beliefs. Delay and uncertainty of outcomes leading to *loss aversion*, overweighting recent events (*availability* heuristic) and long-term memory recall based on a queue of events with memory inhibition of competing material (*associative* heuristic) are all factors that focus attention on salient information (Kahneman and Tversky 2000). Emotions of the individual (*affect* heuristic) and favoritism among options may also play a part in selective attention gathering (Weber and Johnson 2009).

Memory itself is selective and when past experiences are used as decision criteria, perhaps activating the associative heuristic, the context within which these past choices are made acquires importance. The influence and recall of similar past situations make a larger difference when contextualized within "the temporal, spatial and visual context of an experience" (Williams 2018). Moreover, the effect of the reminded context is stronger than the recalled experience itself.[11] This is important because memory recall can be activated by context, even when similar instances have occurred in the very distant past. Therefore, "choices do not always depend upon stable preferences; instead they are constructed dynamically at the time of decision, and critically, via a process that draws on a complex web of contextual association that might be only incidentally related to past decisions of the same kind" (Williams 2018).

Confounding this process, however, is the fact that information itself has several defining characteristics, which could lead to dissimilar decisions under identical information scenarios (discussed in Chap. 3): asynchrony in the way we engage with the world, FOMO or the urgency in being informed in a timely fashion about happenings in the world and ex post valuation or recognizing that information can only be valued after experiencing or internalizing it.

---

of goal-directed behavior. Selective attention operates at very basic levels of perceptual identification. … It also operates at higher cognitive levels, including the initial perception of the situation and assessment of the task at hand (framing, goal elicitation), evidence accumulation … and judgment or choice (determining cut-off or decision rules)" (Kahneman and Tversky 2000).

[11] Bornstein and Norman write "indirect, contextually mediated associations actually exert a stronger effect on subsequent choice than direct associations" (Bornstein and Norman 2017).

This chapter has outlined the four-by-four-by-four approach: four forces of technology, four global facts and four aspects of the sharing model. In the following chapters I show how ICT has shaped risk-taking behavior, how the decoupling of time and content has promoted solo consumption, via texting and streaming videos, and how sharing has stoked anxiety, fear and seclusion. With its multiple facets, sharing can lead to openness and trust as well as fractured communities and nationalism.

These ideas are examined in the following chapters, with Chap. 2 outlining the distinguishing characteristics of information. Chapter 3 explores the underlying motives for entering the social and sharing networks, which feed the advertising model. The aggregate demand for content and the supply of content are examined in Chaps. 4 and 5, respectively, in the context of communication markets. Chapter 6 then traces the arc of this sharing to the individual market for content, while Chap. 7 points the arrow to diminished risk-taking and entrepreneurship.

In Chap. 8, we turn to a more abstract analysis of the broader objectives of connectivity, and Chap. 9 concludes with a vision for the future.

Appendix: Free Will

Attention deficit rests upon not just excess demand for cognitive effort but also a decrease in supply. How is it that mental effort can shrink if we have free will? Can we not decide to remain unaffected by the devices and algorithms? This appendix examines the structure of free will and preferences to clarify the connection between artificial intelligence (devices and algorithms) and mental atrophy.

Recall that we defined free will as agency in the initiation and management of decisions to achieve a desired outcome. Initiation and management of decisions presupposes preferences so we must examine the origins and formation of preferences. This has significance due to the role of preferences in the social contract or the scaffold upon which much of the Internet is based.

The mezzanine floor of this scaffold consists of ideologies and preferences. Upon this foundation we develop decision-making criteria or values for choices and judgment. These criteria get translated into the next level of norms or rules for behavior that govern social ties. These rules are conceptualized in the implicit social contract of trust, reputation, responsibility and rights (TRR&R). The cultural scaffold sustains the socioeconomic network in the long term by becoming tradition, embodied in the law. Thus, it is important to understand preference formation in the first instance.

Implicit in the exchange of digital content is this social contract of trust, reputation, responsibility and rights. Information exchange is based upon an expectation of trust. But it is quite possible that the very act of sharing itself make individuals more trusting. Hence, one can imagine that connectivity and information exchange play a role in the formation of ideologies and preferences. Song lyrics in hip-hop, words in motion pictures, comments on Twitter and images on Facebook can influence our thoughts, even, perhaps instilling new ones. Preferences may, indeed, be endogenous to the system rather than be preordained.

Traditional neoclassical economic theory posits that preferences are *innate*, and that given certain traits, individuals make choices in a utility-maximizing framework. This is the notion of consumer sovereignty, which covers both the desire and ability to choose among multiple options. This view entails free formation of preferences. However, frictions in information provision, such as search costs, and bounded rationality or mental gaps may impact preferences. These gaps and frictions are harnessed by content providers to manipulate our preferences by framing the choices. Changing the context of the choice can change the outcome. Sunstein cites Aesop's tale of the fox and the grapes, in which the fox, gaping at a bunch of delicious grapes, tries jumping to snatch the high-hanging fruit, only to fail repeatedly (Sunstein 2017). Ultimately, he walks off, convincing himself that the grapes were not even worth gaping at. His preferences changed when demand was not met. In a similar vein, sociology views historical and structural features as determining which competing ideology, which set of stories, shared experiences and imagined realities, survives.

Behavioral economics, using psychological process models, maintains that preferences are created to facilitate choices, that psychological processes precede choices and that past experience and the environment are crucial inputs. Rachel Kranton makes a strong case for preferences being endogenous, and not separate from the institutional context. A new framework in the social sciences—identity economics—"considers people in the context of their larger social groups and the motivation those groups engender" (Huettel and Kranton 2012). Choices are made in a context of social and historical structures, which may, in turn, be influenced by these very choices. Individuals may exhibit behavior to align with various groups, to signal that they are of a certain "type" or identity.

A second related question concerns property rights. How do individuals exchange information and pay with cognitive bandwidth hours or attention when there are no clear property rights over personal data or time? Economic theory posits two necessary conditions for markets to function efficiently: property rights and contract enforcement. Without the former, individuals could not trade goods over which they bear no claim, and without the latter, claims have no validity if they cannot be enforced. Drawing upon tradition, claims over personal time are upheld by sociocultural institutions. Economic, social, political and legal institutions embody a system of rules and relationships, an arrangement negotiated by rational, self-interested individuals whose primary goal is maintenance of genetic interests and survival of the species.[12] These institutions embody a system of rules and relationships that provide a way of organizing experience and evaluating reality in order to construct strategies for action.

References

Acemoglu, Daron and Pascual Restrepo. 2018. *Automation and New Tasks: The Implications of the Task Content of Production for Labor Demand*. Working Paper, November 2018 and forthcoming in *The Journal of Economic Perspectives*.

Agrawal, Ajay, Joshua Gans and Avi Goldfarb. 2018. Economic Policy for Artificial Intelligence. In *The Economics of Artificial Intelligence: An Agenda*, ed. Ajay Agrawal, Joshua Gans, and Avi Goldfarb. Chicago: University of Chicago Press.

---

[12] However, Sam Bowles (2016) makes the radical argument that property rights and market structure cannot be perfectly designed to eliminate mischievous behavior on the part of individuals and that an institutional basis of trust and responsibility is vital. Monetary incentives obscure moral incentives and pervert the outcome.

Alon, Titan M., David Berger, Rob Dent, and Benjamin Pugsley. 2018. Older and Slower: The Startup Deficit's Lasting Effects on Aggregate Productivity Growth. *Journal of Monetary Economics*, 93. https://doi.org/10.1016/j.jmoneco.2017.10.004

American College Health Association National Survey of 19,664 college students Accessed on 5/31/19 from https://www.acha.org/documents/ncha/NCHAII_Fall_2018_Undergraduate_Reference_Group_Executive_Summary.pdf

Blinder, Alan. 2018. *Advice and Dissent: Why America Suffers When Economics and Politics Collide*. New York: Basic Books.

Bornstein, Aaron and Kenneth Norman. 2017. Reinstated episodic context guides sampling-based decision for reward. *Nature Neuroscience*. 20 (7), July. Accessed 2/11/19 from https://economics.princeton.edu/wp-content/uploads/2019/02/nn.4573.pdf

Bowles, Sam. 2016. *The Moral Economy: Why Good Incentives Are No Substitute for Good Citizens*. New Haven: Yale University Press.

Brock, William and Lar Peter Hansen. 2018. *Wrestling with Uncertainty in Climate Economic Models*. Working Paper, October 2018. Accessed 2/14/19 from http://larspeterhansen.org/wp-content/uploads/2018/10/brockhansen-new.pdf

Carr, Nicholas. 2008. Is Google Making Us Stupid? What the Internet is Doing to Our Brains. *The Atlantic*, July/August 2008. Accessed 2/11/19 from https://www.theatlantic.com/magazine/archive/2008/07/is-google-making-us-stupid/306868/

Cerqueiro Geraldo, Maria Fabiana Penas, Robert Seamans. November 2018. *Debtor Protection and Business Dynamism*. TILEC Discussion Paper DP 2018-037. Accessed 1/16/2019 from Decker et al. (2018). ISSN 2213-9419 http://ssrn.com/abstract=3212790

Decker, Ryan A., John Haltiwanger, Ron S. Jarmin, and Javier Miranda. 2018. *Changing Business Dynamism and Productivity: Shocks vs. Responsiveness*, Finance and Economics Discussion Series 2018-007. Washington: Board of Governors of the Federal Reserve System. Accessed on 1/16/19 from https://doi.org/10.17016/FEDS.2018.007

Gordon, Robert J. 2016. *The Rise and Fall of American Growth: The U.S. Standard of Living since the Civil War*. Princeton: Princeton University Press.

Griffith, Erin. 2019. More Start-Ups Have an Unfamiliar Message for Venture Capitalists: Get Lost. *New York Times*, January 11, 2019. Accessed 1/13/19 from https://www.nytimes.com/2019/01/11/technology/start-ups-rejecting-venture-capital.html

Gutierrez, German and Thomas Philippon. 2019. Fading Stars. *American Economic Review Papers and Proceedings.* 109 May 2019. DOI: https://doi.org/10.1257/pandp.20191065.

Hamilton, Alexander. *Federalist Papers, #70.*

Huettel, Scott and Rachel Kranton. 2012. Identity Economics and the Brain: Uncovering the Mechanisms of Social Conflict. *Philosophical Transactions of the Royal Society* 367: 680–691.

Kahneman D. and A. Tversky. 2000. *Choices, Values, and Frames.* New York: Cambridge University Press.

Kernighan, Brian. 2018. *Understanding the Digital World.* Princeton: Princeton University Press.

Kornrich, Sabino. 2016. Inequalities in Parental Spending on Young Children: 1972 to 2010. *Sage Journals.* June 8, 2016. Accessed 1/13/2019 from https://journals.sagepub.com/doi/full/10.1177/2332858416644180

Leight, Elias. Producers, Songwriters on How Pop Songs Got So Slow. *Rolling Stone Magazine,* August 15, 2017. Accessed 8/22/2018 from https://www.rollingstone.com/music/music-features/how-did-pop-music-get-so-slow-197794/

López-Salido, David, Jeremy C. Stein, and EgMitchellon Zakrajšek. 2017. "Credit-Market Sentiment and the Business Cycle." *Quarterly Journal of Economics* 132 (3): 1373–1426.

Morris, Ian. 2010. *Why the West Rules – For Now.* New York: Farrar, Strauss and Giroux.

Mullainathan, Sendhil and Eldar Shafir. 2013. *Scarcity: Why Having Too Little Means So Much.* New York: Times Books, Henry Holt and Company, LLC.

Przybylski, Andrew, Kou Murayama, Cody DeHaan, Valerie Gladwell. 2013. Motivational, Emotional, and Behavioral Correlates of Fear of Missing Out. *Computers in Human Behavior* 29(4): 1841–1848.

Shambaugh, Jay, Ryan Nunn, Audrey Breitwieser and Patrick Liu. June 2018. *The State of Competition and Dynamism: Facts About Concentration, Startups and Related Policies.* Brookings. Accessed 4/4/2019 from https://www.brookings.edu/research/the-state-of-competition-and-dynamism-facts-about-concentration-start-ups-and-related-policies/

Sunstein, Cass. 2017. *#Republic: Divided Democracy in the Age of Social Media.* Princeton: Princeton University Press.

Tamir, Diana and Jason Mitchell. 2012. *Disclosing Information about the Self is Intrinsically Rewarding.* Proceedings of the National Academy of Sciences. Accessed 8/9/2018 from http://www.pnas.org/lookup/suppl/doi:10.1073/pnas.1202129109/-/DCSupplemental

Tamir, D.I. and J.P. Mitchell. 2013. Anchoring and Adjustment During Social Inferences. *Journal of Experimental Psychology: General* 142 (1): 151–162.

The Art and Science of Parenting, Special Report on Parenting in *The Economist*, January 3, 2019. Accessed 1/13/2019 from https://www.economist.com/special-report/2019/01/05/the-art-and-science-of-parenting

Vosoughi, Soroush, Deb Roy and Sinan Aral. 2018. The Spread of True and False News Online. *Science.* Accessed 8/22/2018 from http://science.sciencemag.org/content/sci/359/6380/1146.full.pdf

Weber, E.U. and Eric Johnson. 2009. Mindful Judgment and Decision Making. *Annual Review of Psychology* 60: 53–85.

Williams, James. 2018. *Stand Out of Our Light: Freedom and Resistance in the Attention Economy.* Cambridge University Press.

Wu, Tim. 2016. *The Attention Merchants: The Epic Scramble to Get Inside Our Heads.* New York: Random House.

2

Time: The Measure of Connectivity

Digital information and communications technology (ICT) has enabled connectivity on a vast scale, unleashing a market for communication and information. Consuming and processing this information is an activity requiring time and attention. But ICT has advanced by creating an asynchrony in the way we engage with the world. Texting and the mobile calendar are the tools that decouple time and content. By assigning multiple channels of information access to identical slots in our calendar, we are effectively multitasking communication. Text messaging is asynchronous; replies are made at the convenience of the recipient. This chapter explains how and why this kind of tryst with the world matters.

Communication markets have acquired an unrivaled status in today's economic system by virtue of the hundreds, if not thousands or even millions, of connections per individual. Human interaction has moved from face to face to wired (telephone) communication and, more recently, to wireless connections in an ever-expanding network of friends, and platforms for this communication are now commercialized because connections are measured in terms of time. Connections are commoditized, drawing the curtain on the age of transparency. This chapter explains how

© The Author(s) 2019
S. Bhatt, *The Attention Deficit*, https://doi.org/10.1007/978-3-030-21848-5_2

features of information itself lead to this openness. There are six distinctive characteristics of information that standardize, or commoditize, these connections in terms of time.[1]

Asynchrony and Solo Consumption

Asynchrony is the decoupling of time and content with the corollary that individuals have control over time. This is how time becomes a metric of the way we live—our engagement with the world. We connect and obtain information at a speed we determine, optimized for our particular set of preferences. Intensity, or depth, of engagement is correlated to time. Deep engagement requires more time compared with superficial connection. Parallel communication channels, such as open laptops, smartphones on the coffee table and giant computer monitors on the desktop, are examples of asynchrony. Reciprocal conversations with friends may cover intricate elements of one's life, which degenerates into trivia with the presence of a smartphone on the coffee table. It suggests other commitments and that this conversation is not a priority for our friend.

Asynchrony in human interaction, with delayed responses between individuals' texts or photos, expands the network of friends by encouraging parallel text conversations simultaneously. But the likelihood of messages being distorted and misunderstood increases. Face-to-face dialogue may be the most precise form of information transfer since it leaves space for clarification. Letter writing is fully asynchronous since the writer and the reader are physically and temporarily distinct, leaving the greatest room for misunderstanding. The telephone, on the other hand, is geographically distinct, but the conversation is synchronous in real time. Email, and texting, the most popular form of communication between individuals today, is like letter writing.

However, unlike letter writing, where the postal service controls the interval between sent mail and received mail, users themselves control the time delay with email and texting. Agency in delayed communication is

---

[1] Content comprises social communication, factual information exchange, media content and entertainment or Center for International Media Ethics (CIME). Economic literature frequently uses the term information to refer to the entire realm of CIME.

significant because the length of delay itself communicates information. Delayed response time sends a powerful message about the strength of the relationship and significance of the information.[2] Consequently, the intended information leaving person A may not be the same as that received nanoseconds later, in perhaps a different time zone, by recipient B. Further, brevity of texts leaves much room for interpretation.

Asynchrony has given rise to the paradox of solo consumption in a connected world. Earbuds, such as Apple's new AirPods, attached to mobile devices, have made content consumption a convenient and solitary phenomenon. Communication becomes content consumption when the process occurs at a time and pace unrelated to that of others—individuals control when, where and how much content to consume. Compressed production schedules have further decoupled time and content, allowing content providers, such as Netflix, to unbundle content and timing of consumption, enabling solo consumption and binge-watching. The arc of digital communication technology that aimed to connect and facilitate sharing has led to isolation!

Experiments have demonstrated the distraction posed by the presence of devices, a feature of relationships we already intuit. Smartphones hint at multitasking, which "has been shown to decrease concentration and reduce absorption in experiences. In academic environments, media multitasking (e.g. laptop use in classrooms) has been linked to decreases in academic success, presumably because multitasking impairs memory for lecture content" (Ward et al. 2017). Resisting the temptation to check one's phone is itself a "brain drain as limited-capacity attentional resources are recruited to inhibit automatic attention to one's phone and are thus unavailable for engaging with the task at hand" (Tamir et al. 2018). To reduce the mental effort required in dealing with smartphone distraction, individuals gravitate toward heuristics or simplified communication. The availability and affect heuristic might be deployed so conversations made in the presence of smartphones are more likely to become stereotyped or formatted and less improvisational and deep.

---

[2] On the popular comedy show, Key and Peele, Keegan-Michael Key and Jordan Peele eloquently portray the misunderstanding generated by the perceived undertones of their text messages. See https://www.youtube.com/watch?v=naleynXS7yo

Recording friendship by taking selfies on a smartphone is more like "a mnemonic crutch, offloading information onto them and then forgetting that information," signaling the dismissive nature of that interaction—that connection (Tamir et al. 2018). It's as if the relationship is captured, stored and subsequently ignored. Recording and sharing information, in general, diminishes retention due to both multitasking and the mnemonic offloading. It also reduces the general feelings of enjoyment of the experience since sharing leads to comparisons, a notion discussed in Chap. 3. Additionally, there is the element of distraction, which decreases enjoyment directly, as when "using media creates distraction and/or induces mind wandering" (Barnes 2019). The experiments were conducted in a controlled environment as well as in a naturalistic setting. The latter involved comparing participants on self-guided tours who recorded their experience using media with those who used no media. The prediction that memory would be diminished in the former case was validated. The very act of preserving present experiences for the future "prevents people from experiencing them in the first place" (Barasch et al. 2018).

When sharing is the goal, the very experience is undermined. Intentions to share in the future, as with many photo-taking situations, impact a wide range of everyday thoughts and behaviors. Taking photos to share is different from taking photos for one's own recollection, protecting memories, for perusal at a later date. In the former case, individuals are more concerned with how they are perceived by others, which reduces enjoyment. Self-curation of photos is consistent with the notion that people are concerned about their presentation, worried that they may not be seen at their best. "Moreover, self-presentational concern is often associated with pressure to make a good impression and self-conscious emotions such as anxiety" (Barasch et al. 2018). This concern might create disengagement with the experiencing, the living-in-the-moment, itself. In creating a good impression, the focus is on the impression and another point in time—not on current enjoyment. "The intention to share in the future is salient during an experience and negatively affects current enjoyment" (Barasch et al. 2018).

Sharing common resources is laudable except when it becomes a tragedy of the commons. The flame-orange poppies across the hills of Southern California attracted even more visitors to the Super Bloom in

March 2019 after an early Instagram influencer posted selfies amidst the poppy flowers. Over 100,000 visitors were at the town of Lake Elsinore, California, on St Patrick's Day weekend, many of them "off-trail, poppy-picking influencers" who were part of the extensive traffic streaming into the area leading to three-hour traffic jams (Stone 2019).

Choice of Medium

The method of engagement depends upon whether the objective is factual information transmission or social engagement. Texting has become the fastest growing method of communication (Zuckerberg 2019). iMessage, for example, meant for short messages, is perfect for rapid dissemination where time sensitivity of the message takes precedence. The communication is terse and unqualified, often leaving room for misunderstanding of intent. Brevity requires precision in composition of the message with carefully chosen words. Automated spellchecks exacerbate the awkwardness of brief messages when they end up sounding nonsensical.

In a class discussion with freshmen at Princeton University in the spring of 2019, I found that blue and green text message bubbles, which signify iOS and Android phones, have created distinct groups of friends. For the *blues* to get *green* messages signifies a pollution of their message stream and vice versa. The greens frequently resort to Facebook Messenger to avoid the discrimination.

Texting has become the ubiquitous form of communication not only for social purposes but also for business interactions. Small businesses, in particular, are able to communicate in real time with their clients eliminating dead time due to cancelations. A local electrician can cover many more residences by rescheduling via text messaging than waiting for replies to phone messages. This same electrician can examine photos of broken wires sent via text messages prior to arrival, thereby minimizing time spent diagnosing the problem. In all these instances, expediency and time savings have been achieved by communicating the problem in advance via texts.

Posts and updates on social media are longer, and they elaborate upon the main idea. The entire network of one's friends is the audience for such posts, which then opens up space for interpretation, debate and further analysis in the comments section. This interactive yet asynchronous inter-action becomes the most interesting part of many posts. Each comment is time-stamped such that the conversation can be followed as if it occurred in real time—say, at a dinner party. Twitter comments fre-quently convey more depth surrounding an issue than the original post.

Facebook's 2009 manifesto was to give individuals the power to share and make the world more connected and open. But the 2019 version has recognized that the medium is both the "digital equivalent of the living room" with messaging and the "digital equivalent of the town square" with blog posts (Zuckerberg 2019).

Fear-of-Missing-Out

Time is crucial in yet another aspect of information transfer—the fear-of-missing-out or FOMO. Awareness of an occurrence, especially if it involves morbid details, is highly valued and confers a special high status on those who knew this information first. When the Monday morning conversation, or the Sunday family lunch conversation, lands on a recent mishap in town, you would be reluctant, even embarrassed, to admit ignorance. There is an urgency in acquiring the news.

Associated with this phenomenon are powerful network effects, where the value of the information increases with the size of family or friends' network. Essentially, value of information derives from peer esteem—being the first to know of an event accords you an esteemed status in the group and the larger the group, the higher the esteem. Peer esteem derives from group identity or sharing traits with a group (see Chap. 3 for details). Therefore, aligning with a group via access to similar content provides a shared experience and gives individuals an identity and affirmation. It legit-imizes inclusion in the network, whose size increases the value of content.

One of the positive side effects of shared stories is cooperation on a massive scale. These stories become shared interfaces, and just as application programming interfaces (APIs) connect multiple parties across platforms, shared stories are the APIs of the social network. Timely awareness of stories with shared characteristics and traits makes group alignment easier. However, most people are unsure of their traits, their preferences. A could be unsure whether to align with B or some other person simply because A is unsure of its own traits. Access to shared content is reassuring since it helps to define us by group characteristics, group traits and therefore helps to align us with the group.[3] Not only does FOMO drive us to listen to the newest chart-topping hit single by singer Logic, it also brings us closer to his audience, whose characteristics begin to define us.[4]

In the work environment, where information can be time-sensitive and, therefore, critical, not all workers rush to be first in line. In fact, classifying workers as FOMOs and joy-of-missing-out (JOMOs), *The Economist* writes, "Networking events are the kind of thing that gets FOMOs excited as a chance to exchange ideas and make contacts. When JOMOs hear the word 'networking,' they reach for their noise-cancelling headphones. For them, being made to attend an industry cocktail party is rather like being obliged to attend the wedding of someone they barely know; an extended session of social purgatory" (*The Economist* n.d.).

Content Is an Experience Good

Information is an experience good. Products that have a unique fit with different individuals can only be valued after checking them out, much like the menu at restaurants. Despite the high ranking, does the food at a famous restaurant accord with one's tastes and physiology?

Familiarity, over time, with a particular content platform increases its value and you then desire to consume more content from that same source. Costs of switching to a new platform are higher due to this stickiness and media platforms can raise prices or, more commonly, add features which

---

[3] Violence and acrimony that has been recently enveloping the nation could be attributed to generally disenchanted individuals aligning with groups in search of identity and legitimacy. They then acquire the most extreme views of that group.

[4] Logic's hit single 1-800-273-8255 was third on the Billboard Hot 100 list in 2017. The song's name is the phone number of the National Suicide Prevention Hotline. Calls to that number increased significantly, following the song's release as well as after the night of the 2017 MTV Video Music Awards. Did the song bring awareness of suicidal tendencies among teens or did it simply provide an outlet for help?

Facebook, YouTube continue to be the most widely used online platforms among U.S. adults

% of U.S. adults who say they ever use the following online platforms or messaging apps online or on their cellphone

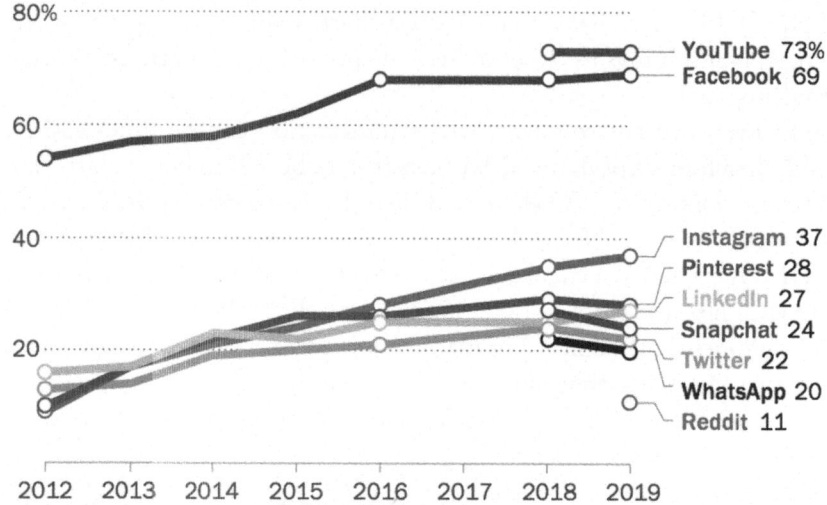

Note: Pre-2018 telephone poll data is not available for YouTube, Snapchat and WhatsApp. Comparable trend data is not available for Reddit.
Source: Survey conducted Jan. 8-Feb. 7, 2019.

PEW RESEARCH CENTER

Fig. 2.1 Majority of Americans now use Facebook and YouTube. (Source: Share of adults using social media is unchanged since 2018, Pew Research Center April 10, 2019. Retrieved on 6/5/2019 from https://www.pewresearch.org/fact-tank/2019/04/10/share-of-u-s-adults-using-social-media-including-facebook-is-mostly-unchanged-since-2018/)

extract valuable personal details without any decrease in subscribers to the platform. Driven by competition from Snap in 2015, Facebook started offering short video clips as part of its slate of tools. Figure 2.1 shows that Facebook users comprised over 50 percent of US adults in 2012 before flattening out to 68 percent in 2016–2018, suggesting that users were becoming suspicious of the surfeit of possibly nefarious tools.

Asynchrony, choice of medium, FOMO and familiarity with platform are time-sensitive attributes of information. Time-invariant characteristics of information are its public good nature and its distinctive cost structure.

Information Is a Public Good

Information is non-excludable and non-rival—a classic definition of a public good. "Information is costly to produce but cheap to reproduce" and hence it is difficult to exclude people (Shapiro and Varian 1999). Information is non-rival because any two people can simultaneously consume the good. Persons A and B can both consume the very same piece of information, unlike a good like a car, which cannot be driven by multiple persons. In fact, it can spread nearly instantaneously across the entire network: this is the logic behind some pieces of information going viral!

It is precisely these two features of markets that necessitate an underpinning of trust. How can we write property rights on particular pieces of digital content and how can we enforce these rights? Copyrights on digital content are difficult to implement and originality is difficult to verify. Current law is more than 40 years old and merely offers protection (under title 17 of USC, which includes the Copyright Act of 1976) to authors of original works of authorship that are fixed in a tangible form of expression. A work is fixed when it is captured in a sufficiently permanent medium (Copyright Law of the United States 1976). As we shall see in Chap. 5, intellectual property is the most valuable asset of entertainment companies.

Non-rivalry and non-exclusivity means that "nothing stays totally secret in the social media age—even deleted tweets" (Barnes 2019). In the world of politics and the entertainment industry, where individual personalities have millions of votes or dollars riding on their name, there are two powerful strands coming together: a "cultural intolerance for harassment and bias; and the accessibility of everything, from a decade of Twitter posts to videos taken at high school parties to college term papers," according to a partner at one of Hollywood's leading media relations agencies, Principal Communications (Barnes 2019). Hiring intelligence experts who can do background checks can buttress the good name of critical persons. Effectively, we are in a world with stricter rules and enhanced enforcement, both driven by the public good nature of information.

As a result, underinvestment in content may arise when the creator cannot capture all the costs of producing content. To be sure, there is the superstar world where some creative artists can successfully monetize their talent. In the audio world, Chance the Rapper has built his career on SoundCloud, the user-created music streaming platform, with 1.5 million followers and a true talent in bringing together his roots in rap, hip-hop, jazz and gospel music in formidable rap lyrics.

Underinvestment and Overuse of Information

The public good nature of information leads to the common resource problem: underinvestment and overuse. Content production is costly and best amortized by standardization and homogeneity. More precisely, production costs of digital content are large and non-recoverable, while the marginal costs of distribution are near zero and so there is enormous scalability. The addition of one more user costs no more to the creator and distributor, reinforcing the formation of media behemoths. The response of content creators, both user-generated and commercially generated, to the high fixed costs of content creation and the near-zero costs of distribution is to provide homogeneous or weakly differentiated quality. Content is now limited to popular topics, with popularity itself being self-reinforcing as people gravitate to the top-ranked trending topic.

One of the consequences of this narrowing of exposure is in the realm of citizenship and "deliberative democracy." According to Sunstein (2017),

> to the extent that both citizens and representatives are acting on the basis of diverse encounters and experiences, and benefiting from heterogeneity, they are behaving in accordance with the highest ideals of the constitutional design.

But he goes on to warn that

> When people deliberate together, they often give disproportionate weight to "common knowledge"—information that they all share in advance. By contrast, they frequently give too little weight to unshared information—information that is held by one or only a few people. There is every reason to think that the same asymmetry is occurring online.

Curated content, as in *The New York Times* or *The Washington Post*, is costly to produce, necessitating the imposition of subscription fees. This feature promotes allocation of content by income, making it a luxury good. In other words, when incomes rise, the consumption of luxury goods rises in equal or greater proportion.

In the video world, technologies such as virtual reality and augmented reality have added to the fixed costs of production, resulting in increasing *returns to scale in production technology*. With plentiful supplies of inputs, the average costs of content production will decrease as quantity of content increases, allowing firms to exploit *economies of scale*, which applies to overall supply. As a consequence, organizational behemoths, like Amazon and Netflix, arise since only large firms can sustain this sort of cost structure. The recent spate of mergers in the entertainment industry points to exactly this phenomenon.

What does this imply for content quality? In addition to the issue of underinvestment in content, generally (the external margin of content), quality adds another dimension to content production (the internal margin of content). If superior quality entails higher fixed costs, then average costs will be higher, and a larger audience will be needed to amortize these costs. Quality is contextual, so what qualifies as high quality for a New Yorker would not necessarily be regarded as such by a Chinese viewer. Consequently, the large audience for a particular genre evaporates when diverse populations' tastes must be met. A generic formula is then applied to all content, as in the case of franchise films or the Marvel Cinematic Universe (MCU):

> The franchise film era is, in many ways, a return to the studio system. Only now the major entertainment companies don't own the most important

talent—they own the most important cinematic brands. Instead of fighting for a deal at MGM or Paramount, actors and filmmakers vie for a chance to make the latest spinoff of Star Wars or X-men. (Fritz 2018)

The American studio world has swung away from drama and dialogue toward the visually charged and "high-concept" film. "High concepts, in general, favor grand situations, exciting action and stimulating visuals" over character development and scintillating banter, so they can be summarized in few words (Stern 2019).

In the audio world, song production is dominated by the digital audio workstation (DAW) and a keyboard, headphones and microphone. Music producers mix and match the vocals, instrumentals, drums and bass lines, which are all kept in separate files or stems. Even if there aren't separate files, clever use of the DAW will allow chopping off various stems. Once the beats-per-minute (BPM) are decided and melody added, the song is remixed and ready to be uploaded on SoundCloud for free. Posting remixes of famous artists, collaborating with these artists, cross-promoting on Twitter and creating a website are the trendy routes to attaining chart-topping hits. Music industry connections and merchandising also help marketing on SoundCloud. However, is the song original only if the melody and lyrics are distinct or is the remix an infringement of copyright?

Content Determines Content

The issue of consumer sovereignty, raised in Appendix 1, also pertains to content choice. How does content itself determine preferential selection of future content? The images we view, the music we listen to and the words we read are imperceptibly blended into our consciousness, and they influence our future preferences. Content consists of our shared stories, experiences and common myths, and it becomes the underlying ideology, the imagined reality that binds together larger numbers of strangers. Values that form the criteria for judgment and ranking of alternatives are based on these ideas. Harari writes, "Imagined reality is something that everyone believes in, and as long as this communal belief persists, the imagined reality exerts force in the world" (Harare 2017). So, content influences our beliefs.

Making choices on the basis of recommendations and rankings is effectively a process of delegating decision-making to the crowd or the curator. A film rated five stars on Netflix will encourage more viewers effectively catapulting it into the highest rank. Producers, noting this process, will design future movies with similar traits, so popularity drives future content. The proliferation of Hollywood sequels is testimony to this feedback process, as is discussed in Chap. 5.

Can popular sentiment on social media, then, impact consequential decisions? Do social media behemoths have the power to shape popular culture? Legal decisions, voting decisions and economic decisions are all based on a combination of public perception, available information and judgment. Judgment consists of a ranking of choices, which depends upon the criteria chosen. When digital content impacts preferences and beliefs, then the very criteria for judgment, ideas, are changed. This makes content markets a vital conduit for decision-making. When popular sentiment becomes hardened into core ideas, judgment is, indeed, impacted.

Public sentiment aggregates independent unbiased opinions as well as biases induced by social influence. Racial stereotypes are, in some sense, founded upon mixing private information with commonly held beliefs. Suppose a jury has to pass judgment on a prisoner. Will the jury be imperceptibly swayed by social media–driven sentiment, by generally held stereotypes?

Should jury members minimize a Type I error, which is the risk of falsely proclaiming guilt, when the prisoner is innocent? How do we evaluate the consequences of penalizing a potentially innocent party? On the other hand, Type II error is the risk of setting free a defendant accused of a serious crime. How do we evaluate the dangers of an alleged psychopath set free on the streets?

Stereotypes play a meaningful role in bail decisions. Dobbie et al. show, using 2010–2014 data from Miami and Philadelphia, that "bail judges are racially biased against black defendants, with substantially more racial bias amongst inexperienced and part-time judges" (Arnold et al. 2018). Dobbie tested for pre-trial misconduct rates (jumping bail) for newly incarcerated black and white defendants and found that the rate of misconduct is higher among white defendants as they are more likely to be released, while guilty than otherwise equivalent black defendants. The bias is significantly higher among the cohort of judges who are more likely to be guided by public sentiment—inexperienced and part-time judges that rely on inaccurate stereotypes as they search for resources to help them evaluate defendants. The relative danger of releasing a black defendant is stereotypically assumed to be higher by these judges. The authors find that in addition to simple statistical correlation between observables traits such as race and unobservable data such as crime rates, there is evidence for racially based prediction errors in risk. They find "that there is a 23.6 percent rate of rearrest for marginally released white defendants and a 1.4 percent rate of rearrest for marginally released black defendants" (Arnold et al. 2018).

Fear of public outrage, as fingers point to well-known stereotypes, could be the tipping point for judges as they maximize Type I error, the risk of falsely accusing an innocent black defendant, and minimize Type II error, the risk of setting a guilty black man free.

Consequently, we are at a crossroads of technology, law and human behavior. In order to reconsider the law and regulation, we need to understand the impact of digital content on human behavior. The question facing us now, ever more urgently, is how we can assure that technology enriches the human experience by supporting creativity, entrepreneurship and imagination while simultaneously preserving our fundamental values, such as personal privacy and national security. I examine these issues in Chap. 8. But first we need to understand communication markets, how content is created by commoditizing connectivity in the sharing model and how this content then loops back to influence human behavior, which is the subject of Chap. 3.

References

Arnold, David, Will Dobbie and Crystal Yang. 2018. Racial Bias in Bail Decisions. *The Quarterly Journal of Economics* 133 (2), November 2018. https://doi.org/10.1093/qje/qjy012

Barasch, Alixandra, Gal Zauberman, Kristin Diehl. 2018. How the Intention to Share Can Undermine Enjoyment: Photo-Taking Goals and Evaluation of Experiences, *Journal of Consumer Research* 44 (6), April, pp. 1220–1237, https://doi.org/10.1093/jcr/ucx112

Barnes, Brooke, NYT February 4, 2019. *Scouring Hollywood's Background, Before Someone Else Gets There First.* Accessed 2/4/19 from https://www.nytimes.com/2019/02/04/business/media/-celebrity-backgrounds-media-relations.html?action=click&module=Well&pgtype=Homepage§ion=Business

Copyright Law of the United States. 1976. Accessed 9/3/2018 from https://www.copyright.gov/title17/

Fritz, Ben. 2018. *The Big Picture: The Fight for the Future of Movies.* New York: Houghton Mifflin.

Harare, Yuval Noah. 2017. *Sapiens: A Brief History of Humankind* and *Homo Deus: A Brief History of Tomorrow.* New York: Harper Collins.

Shapiro, Carl and Hal Varian. 1999. *Information Rules.* Boston: Harvard Business School Press. P 3.

Stern, Elias. 2019. *High Concept: An Empirical Analysis of Globalization and the Cultural Discount Effect on the American Motion Picture Industry.* Senior Thesis presented to the Department of Economics, Princeton University, 2019.

Stone, Zak. 3/23/2019. *Under the Influence of a Super Bloom.* Accessed 3/29/2019 from https://www.nytimes.com/2019/03/23/style/super-bloom-california-instagram-influencer.html?algo=als1&cmpid=73&module=newsletter-best-reads&nl=personalization&nlid=20718461&rank=2&recid=1J8QTWOmbnzcotGsFW7vbXaLnWh

Sunstein, Cass. 2017. *#Republic: Divided Democracy in the Age of Social Media.* Princeton: Princeton University Press.

Tamir, Diana, Emma Templeton, Adrian Ward, Jamil Zaki. 2018. Media Usage Diminished Memory for Experiences. *Journal of Experimental Psychology* 76.

The Economist, The Two Tribes of Working Life. Accessed 2/5/2019 from https://www.economist.com/business/2019/01/31/the-two-tribes-of-working-life

Ward, Adrian F., Kristen Duke, Ayelet Gneezy, and Maarten W. Bos. 2017. Brain Drain: The Mere Presence of One's Own Smartphone Reduces Available Cognitive Capacity. *Journal of the Association for Consumer Research* 2 (2), April 2017. Accessed 3/28/19 from https://www.journals.uchicago.edu/doi/10.1086/691462

Zuckerberg, Mark. 2019. *Manifesto.* Accessed 5/22/2019 from https://www.facebook.com/notes/mark-zuckerberg/a-privacy-focused-vision-for-social-networking/10156700570096634/

3

The Psychology of Connectivity: Follower Counts and Identity

How did we get from a world of universal connectivity to a world of powerful content? Why do we share? Optimization of connectivity, by maximizing the number of followers on social media for example, is the sharing model. Maximization of followers entails financial rewards, psychological rewards of self-disclosure and an affirmation of identity and hence of self-esteem. All three reward systems drive social media content. Further, each connection generates dynamics, which sets in motion a further search for connections. We begin with a discussion of the rewards—financial and psychological—before elaborating on identity.

Financial Rewards

Optimization of an outcome with respect to product choice first requires standardization of the product. Optimizing rewards or utility with respect to connectivity entails standardization of connectivity, which means framing it as a commodity. This allows us to extract the highest value from connectivity.

Communication markets have commoditized connectivity by assigning values to "friends." Examples include metrics such as the number of Facebook followers (and likes), of Instagram followers, of Twitter followers

© The Author(s) 2019
S. Bhatt, *The Attention Deficit*, https://doi.org/10.1007/978-3-030-21848-5_3

(and retweets), the size of subreddit communities, which make up the backbone of Reddit.com, and the social news site.[1] In fact, follower counts on social networks determine labor market outcomes, such as salary and job description. On Twitter, the ratio of the number of replies to the number of likes and retweets is meaningful—a high ratio is bad news, since replies are addressed only to the author of the tweet, whereas likes and retweets are broadcast more widely. High follower counts are important for influencers or amateur tastemakers and YouTube stars, where advertisers spend billions on sponsorships (Confessore et al. 2018). A sponsored post on Instagram for two-year-old identical twins with over 2 million followers on Instagram earns them between $10,000 and $20,000 (Maheshwari 2019).

Naturally, content, and brands, that earn more "friends" rises to the top of charts.

The National Basketball League has the youngest television audience among all televised sports due to its masterful use of social media. When the Golden State Warriors beat the Oklahoma City Thunder in the spring of 2018, there was a torrent of words on Twitter, with the Cleveland Cavaliers' LeBron James himself tweeting on February 28, 2018: "@StephenCurry30 needs to stop it man!! He's ridiculous man! Never before seen someone like him in the history of ball" (McClusky n.d.). Three months later, Curry's team beat LeBron James' Cavaliers! "The first thing that players do when they walk off the floor and head for the locker room," Twitter COO Adam Bain says, "before jumping in the shower or anything, is grab their phones and fire up Twitter to see all of the public reaction and conversations about how they performed that night" (McClusky n.d.). In a study commissioned by Facebook, the authors found an additional post in the 15 minutes prior to game-time correlated with 1000 new viewers in the initial broadcast minute. "That's not a threat to traditional broadcast; it's a lifeline" (McClusky n.d.).

In this influence economy, these numbers can translate into personal wealth:

---

[1] Twitter follower counts for Nobel prize-winning economists vary from 4.52 million for Paul Krugman to 247,000 for Joseph Stiglitz and 140,000 for Richard Thaler. Numbers for academic economists who write extensively on urgent policy issues such as trade also vary from 118,000 for Dani Rodrik and 116,000 for Kaushik Basu to 13,000 for Ken Rogoff. See IDEAS (2019) for details.

The world's collective yearning for connection has not only reshaped the Fortune 500 and upended the advertising industry but also created a new status market: the number of people who follow, like or "friend" you. For some entertainers and entrepreneurs, this virtual status is a real-world currency. Follower counts on social networks help determine who will hire them, how much they are paid for bookings and endorsements, even how potential customers evaluate their businesses or products. (Confessore et al. 2018)

When brand promoters' earnings, as an influencer on Twitter, is a function of the number of followers, there will arise a market for this product. In fact, there is a thriving global market for fake followers, called bots, which are based on dormant accounts on Twitter. An economic forecaster purchased followers from a commercial site, Devumi, to boost his credibility (Confessore et al. 2018)! On the other hand, Elon Musk's infamous tweet "Am considering taking Tesla private at $420. Funding secured" on August 7, 2018, cost him the chairmanship of Tesla for three years and a $20 million fine. According to the Securities and Exchange Commission, Musk made "false and misleading claims" about the company already having the funding to go private (SEC Press Release 2018).

High follower counts can be a signal for competence or simply for popularity and become self-reinforcing when fans allocate their time according to follower counts. If new users follow those with the highest follower ranking, the count for the top-ranked celebrities spirals upward. However, the high follower count does not reflect engagement with the content of the tweet, which is better represented by retweeting. Another metric, called "reach," which is the follower count of followers, is used by researchers to show that by actively engaging a large number of Twitter followers, scientists can "increase the odds of being followed by a decision-maker who might see their messages, as well as the odds of being identified as a potential expert for further contributions" (Côté and Darling 2018). While the average academic scientist with fewer than 1000 Twitter followers is followed mostly by other scientists, those above this threshold connect with a more diverse audience. Whether high follower count itself, a higher reach or high retweeting percentage is a more accurate predictor of influence remains an important research question (Côté and Darling 2018).

YouTube has been the launching pad for a number of successful peer-to-peer businesses, built on an initial YouTube video. Budding musicians, would-be lead singers for a band, aspiring cosmeticians and makeup artists post elaborate videos of their skills on YouTube. Many gather enormous followings that jump-start their own successful firms. Ipsy, for example, was started by Michelle Phan in 2007 when she posted YouTube videos, or video blogs, that were short beauty tutorials. It was formally launched in 2011 as myglam.com, and later in 2012 as Ipsy.com, which followed the subscription model (Chap. 5 explains this model). For a fixed monthly fee, users receive a bag of cosmetics samples, which are provided by the firms as free marketing tools. Their value in this exchange lies in the feedback received from the roughly three million subscribers. Today, Ipsy gives vloggers access to a 10,000 square foot space in Santa Monica where they produce videos and then broadcast online (Phan 2007).

Psychological Rewards

For the average social media user, high follower counts, in the hundreds rather than in the millions one associates with celebrities, generates a feeling of well-being similar to that obtained from food or sex. However, the suggestion of a vast network of friends is, in fact, misleading. Imperfect processing of friendship ties and comparison across this leads to a spiral of ever-shrinking self-esteem. There are two reasons why the initial gratification from being followed by large numbers of individuals diminishes due to the ensuing negative emotions.

First, while digital communication technology has facilitated a vast number of connections, most of these are beyond what can be properly processed by individuals, since the basic human ability to maintain ties reaches a maximum at 150 links. Yuval Harare (2017) feels that

Humans, like chimps, have social instincts that enabled our ancestors to form friendships and hierarchies, and to hunt or fight together. However, like the social instincts of chimps, those of humans were adapted only for small intimate groups. When the group grew too large, its social order destabilized and the band split. Even if a particularly fertile valley could feed 500 archaic Sapiens, there was no way that so many strangers could live together. How could they agree who should be leader, who should hunt where, or who should mate with whom. [loc505]

When individuals have more friends, they have more options as to whom to spend time with. However, since time is fixed, either one has fewer friends or less time is spent with each friend. Optimizing friend interaction with respect to time requires precise calibration of rewards under various scenarios, which can be stressful. Once again, time is the critical decision variable. On the one hand, meaningful relationships ensuing from conversational backroads that reveal personality cannot be created with a time constraint hovering in the background. On the other hand, short interactions, as in speed-dating, can provide the opening for a relationship.

In addition to the vast numbers of connections facilitated by ICT, imperfect processing of friendships can also arise due to multiple formats for interaction. Communication is synchronous, which can be face to face or via the telephone, or asynchronous, in the form of digital texts and images. Should time be spent texting or meeting in person? Just as inflation puts relative prices into flux, ICT puts relative relationships into question. If A texts B more often, but has infrequent, longer face-to-face meetings with C, how do B and C rank in A's esteem. Virtual or digital connections are sequential and necessarily less spontaneous since there is a time delay, while real connections are simultaneous since physical presence conveys messages via body language, which substitutes for dialogue. Timing and format of connections set the criteria for an implicit ranking of friendships, a ranking which can be unsettling for all and leads to what I discuss later as despair.

Second, friendship and social connection entail sharing and, therefore, comparison. For example, after a day of fly-fishing with a friend, we share and necessarily compare the size of fish we caught. Beyond good-natured comparison, there lurks envy. There is a slippery slope leading from joy due to a friend's achievements to wanting what they have (jealousy), or worse, recognizing that I can never match their success (envy). An overwhelmingly large social network could easily slide into too much sharing and then, inevitably, multiple comparisons.

Sharing, therefore, is actually composed of two distinct elements: *sharing* itself and *comparing*. Enveloped in a mindset of transparency and trust, individuals share, and implicit in this openness is risk-taking. There are no guarantees about the outcome of this sharing. There is simply faith in the system. *Comparing*, on the other hand, opens the door to jealousy, envy, resentment and, ultimately, a more universal emotion—fear. There is a perception of danger, real or imagined, in the world and with it comes risk aversion. The response is to create walled gardens of privacy. Effectively, sharing is a risky enterprise, necessitating trust, while comparing is a risk avoidance exercise to guard against being knocked out of balance in the social realm.

Why do friends share? What is it about social connections that compel us to publicize the most intimate details of our lives? What, exactly, are the psychological rewards of sharing? Shared experiences confer a sense of belonging and acknowledgment of one's existence or legitimacy, which gives meaning to the human experience.

Sharing information about dangers lurking in the deep woods, or group dynamics, is a survival strategy. This kind of external sharing is about the environment, about aspects of our lives that have common impact.

> Social cooperation is our key for survival and reproduction. It is not enough for individual men and women to know the whereabouts of lions and bison. It's much more important for them to know who in their band hates whom, who is sleeping with whom, who is honest, and who is a cheat. … Reliable information about who could be trusted meant that small bands could expand into larger bands, and Sapiens could develop tighter and more sophisticated types of cooperation. (Harare 2017)

On the other hand, self-disclosure, or internal sharing, is about oneself, one's own possessions and successes or failures. This kind of sharing is a signal to the world: *I exist and here is who I am.* Self-disclosure validates one's existence. It provides a place of legitimacy, a secure benchmark, in an uncertain world. According to psychologists Tamir and Mitchell, "self-disclosure was strongly associated with increased activation of brain regions that form the mesolimbic dopamine system" (Tamir and Mitchell 2012). This affirmation of one's identity activates regions of the brain commonly associated with food and sex. Self-revelation provides the security of knowing that one's survival is assured by articulating one's traits.

Affirmation is also elicited by sharing experiences. In a brilliant recent experiment, Diana Tamir et al. find that people will pay to watch movies simultaneously with others despite viewing the screen in separate rooms. The very idea of sharing anything generates connection, which is hard to define, but it is easy to measure using brain activity in fMRI scans.[2]

Neuroscience studies of decision-making and social cognition capture the underlying motivation for present bias in rewards and identity, respectively. Using magnetic resonance imaging scanners, researchers are able to locate the regions of the brain activated by delayed rewards, cooperation and food—the brain regions activated in all three cases lie in the prefrontal and frontal cortex. Identifying and sharing with a group, generating self-esteem also appears to be linked to the frontal cortex (Huettel and Kranton 2012).

In addition, not merely sharing, but cooperating with a group activates this region. An early neuroimaging study of the repeated Prisoner's Dilemma game, where two prisoners can either cooperate to get the standard prison sentence or snitch on the other to attempt a more lenient sentence, found "that signals of ongoing cooperation led to increased activation in reward regions," the same region activated by food and sex. Furthermore, "[a]ctions that increase the fairness of resource distribution, compared with an equally rewarding outcome (for oneself) that does not reduce inequity, lead to greater activation in the reward system" (Huettel and Kranton 2012).

Blogging, or reaching out to friends via text, is one form of sharing. Another form is posting photos on Instagram together with a brief message. In the first case, for example, by sharing thoughts on Reddit, I am broadcasting my thoughts and views to an audience interested in the idea rather than the person disseminating the idea. In the second case, I am

---

[2] Tamir et al., unpublished working paper, Department of Psychology, Princeton University, 2019.

publicizing either my true image or a disguised image reflecting an ideal or aspiration. The image aligns various characteristics that describe personality with my perception of the average character of the aspired social network in order to receive affirmation. You need the groups' acknowledgment to justify your legitimacy "Am I thinking the right thoughts?" "Am I doing the right thing?" How do I stand relative to the group? By showing pictures of you marching in Washington D.C., with the hashtag #MarchForOurLives, you are signaling that you too support gun control. By uploading text to the hashtags #MeToo and #MeTooMVMT, you are signaling that you support people who have been sexually abused or that you are on the same page as followers of #TimesUp.

To the extent that personality traits exhibit *intersectionality*, a picture can be worth a thousand words, as explaining a person's unique combination of race, ethnicity, history and culture may not be amenable to clickbait conventions of abbreviation and syntax. When pictures are used to show off, to signal status by uploading images that outline a fun-filled life, devoid of the hustle and bustle of daily penance, the entertainment aspect of sharing supersedes the informative aspect, it becomes a fantasy of stories about lives lived in an imaginary world. But this is precisely where the stories click negative feelings of envy and jealousy when comparing one's own life with that of the narratives on social media.

Identity and Self-esteem

In the psychological model, self-revelation and identity are about physiological changes induced by dopamine spikes. In the economists' view, sharing is about aligning with groups in order to build peer esteem and judging relative status with respect to other individuals to acquire self-esteem. Decisions are based on preferences and behavioral norms—people care about how others behave. Moreover, since this behavior is contextual in the sense of who is doing the interacting, we define social categories and "the term identity is used to describe a person's social category" (Akerlof and Kranton 2000). Akerlof and Kranton posit that "because identity is fundamental to behavior, choice of identity may be the most important economic decision people make. Individuals may—more or less consciously—choose who they want to be" (Akerlof and Kranton 2005).[3] This choice can be thought of as framing the environment (Akerlof and Kranton 2000).

By sharing, individuals align with groups for affirmation and build esteem by comparison with other groups. Unpacking behavior into a set of traits or characteristics, one can further define identity in terms of traits rather than behavior, so a person of a certain type O, shares traits with all others of type O. All individuals with common traits are judged equally and if they share traits they are deemed to belong to the same group. This line of thinking assumes that there is a unique correspondence between behavior and traits—that one can infer identity from actions. When people with similar traits form a group, they are more likely to coalesce around a common goal and behave according to norms particular to that group.

By posting a picture reflecting a social category, a person's self-image is established. The picture is associated with actual or fictional persons who model ideal behavior and any suggestion of deviation from this behavior imposes anxiety on individuals themselves as well as members of their group.

However, in a world where self-perception is weak or lacking, David Akerlof (2016) proposes a theory where people define their identity. Characterizing individuals by their personality, skills, race, gender and other traits, he writes "I propose to define personal identity as one's belief about one's type. Or, put another way, personal identity is who one thinks one is" (Akerlof 2016). Building upon this idea, one can define self-esteem in terms of these traits so that higher self-esteem requires a belief that all people don't share the same traits. In fact, we "are normally inclined to disesteem one another since only by esteeming others less can one raise one's self-esteem; but, when [we] identify with one another, it creates incentives to positively esteem one another; there is a mutual desire to judge the shared type well" (Akerlof 2016).

---

[3] Individuals choose actions in the short term consistent with certain identities, but they can choose identity itself, as well as change the behavioral norms consistent with that identity, in the long term (Kranton 2016).

The notion of fitting in is a way of building peer esteem. Self-esteem and peer esteem are the result of relative judgments or norms about traits and behavior. Norms set up the benchmark for what is good or bad behavior in various situations. Then self/peer esteem follows from "a judgment of how good an agent/group is relative to a comparison population" (Akerlof 2016). When an individual "thinks in we-terms," she becomes cognizant of the esteem accorded to the group rather than to her so that, subsequently, her choice of characteristics and actions place her in alignment with the group. It is quite possible that a person acts in accordance with the norms for a particular group prior to being accepted into the in-group. Motivation for collective action toward a common goal arises from being accepted into the in-group, for example, so that we can "define we thinking as a mode of thinking in which an individual takes a group's goals as his own" (Akerlof 2016).

Identifying with a group and judging one's group to be superior leads to positive peer esteem. On the other hand, disidentification, or separating oneself from the group, is one way of increasing self-esteem (Akerlof 2016). So, while one's desire for affirmation and acknowledgment leads to sharing with groups, it is also the case that our desire for esteem leads to comparisons. I can only be of higher status if I have a different, superior set of traits. Consequently, sharing is a highly subtle art: By posting images or text online, I want to illustrate common traits to establish group identity but, simultaneously, signal differences for establishing superiority and heightening self-esteem. Sharing vacation images of hiking in Patagonia with multiple hashtags on Instagram, for example, signals your identity with #Patagonia groups and establishes peer esteem. But images of you at the top of Mt. Fitzroy in Argentina with hashtag #Fitzroy raise your self-esteem by distancing yourself from the general crowd of trekkers.[4]

This line of thinking leads to comparisons and judgment on social media. Failing to measure up to conventional norms leads to feelings of

---

[4] Akerlof writes "New Yorkers enjoy, for example, being able to engage in mutual pride-taking over a win by the Yankees. Therefore, just as agents' desire for self-esteem motivates disidentification with Boston [Red Sox], the desire for peer esteem motivates New Yorkers to forge a common group identity" (Akerlof and Kranton 2000).

envy and distress. Will my lack of cultural embeddedness jeopardize my social embeddedness? Goldberg et al. write:

> The term cultural embeddedness refers to the degree an individual internal-izes the common culture and accepts group norms. It describes the extent to which that person shares values and assumptions with those around her and how much the common culture shapes her interactions with others. (Goldberg et al. 2016)

Sharing and comparisons sow discord among competing groups. For example, college-educated members of minority groups threaten to devalue non-college educated members of their group; the latter then ridicule and diminish the achievements of the high achievers. Another example comes from crossing gender-based occupations when women working in predominantly male occupations lower men's self-esteem by making them feel less masculine (Akerlof and Kranton 2005). The in-group and out-group differences are accentuated by sharing.

In the work environment, "employees who are good fits are more satis-fied, more strongly attached, better motivated and better performers than their peers who are not in sync culturally" (Akerlof and Kranton 2000). The analogy with a social environment is even more seductive. Peer esteem is strongly connected to *fitting in*. Friends who don't *fit in* with the group ideology risk their very group membership. Hence, the natural reaction is further interaction on social media, perhaps to recalculate the discrepancy between one's personality and ideas and those of the group. This perpetuates a spiral of even more connections with more compari-sons and consequently more distress.

The need to belong is a powerful motivation for online groups to orga-nize around some focal emotion, such as hate, as a kind of performance in solidarity with the group and its values. There is a sense of strength and belonging in this sort of performance. Performing according to the norms of the group by publicly espousing group values through words and, quite possibly violent action, is a way to strengthen ties to the in-group and prove that they, indeed, are members in good standing. Attacking perceived outsiders to the group signals loyalty and differentiates in-group from out-group members. Moreover, in online platforms influential figures are precisely those who post most often and, therefore, whose voice is most heard (Gomila et al. in prep).

In some situations, it may be optimal for an individual to switch groups if changing identity is cheap and if fitting in with the socially given group is costly. Costs in terms of emotional conflict for oneself and for other group members may be incurred when observing own-group norms and behavior. Then crossing over to another group and switching identities may be the solution. In an online series, a reporter at the Chicago Tribune writes:

> Lesbian. Gay. Pansexual. Queer. Black. We use various terms during conversations about sexuality and gender expression, but regardless of our words, so many of these discussion boil down to identity. When you belong to two distinctive groups—and two marginalized groups, at that—these distinct qualities weave and intersect even as they operate on parallel track within the dating scene.
>
> … As one black and queer interviewee puts it, "I try to use the word identity very often. These are my identities, but they are not my labels." Hence, shaping one's personality is moving away from distinct labels. (Carpenter 2018)

Evocative Images and Violence

Connections can be forged in multiple ways, as explained in Chap. 2: face to face, voice (over telephone), text and images. Snap has bet its future on the latter form of communication, or visual communication. Sharing with images is more efficient, since we think in terms of images. According to Tony Damasio, "Having a mind means that an organism forms neural representations which can become images, be manipulated in a process called thought, and eventually influence behavior by helping predict the future, plan accordingly and choose the next action" (Damasio 2016). So according to this view of human consciousness, images prevailed first and these were later encrypted, using language, into thoughts. According to Damasio, Descartes' famous line "I think therefore I am" is Descartes' Error since we visualize before we think. Perhaps, the line should be rephrased to "I share therefore I am."

Images, in some cases, acquire destructive power. Positive peer esteem that derives from groups coalescing around an inflammable topic can be life threatening. Posts that tap on some latent grievance or past history are like church bells calling for group bonding over vigilante tactics. Furthermore, connectivity via social media amplifies basic human instincts that lead to, for example, lynching and yet more violence. Notions of democracy and equality are defaced by suggestive rumors (Fisher and Taub 2018).

Facebook's newsfeed, for example, is based on an algorithm which prioritizes "content that wins the most engagement, [whatever keeps the user on the site the longest]. We know from studies that negative, primal emotions—fear, anger—draw the most engagement. Posts that provoke those emotions rise naturally" to the top. Also, "posts that indulge your group identity [tribalism] by attacking another group tend to perform really well. … And nothing delivers a dopamine hit like posting something that will draw out tendencies toward angry, fearful tribalism" (Fisher and Taub 2018).

Suppose Facebook's goal is to maximize the number of hours you spend on their site. If "likes" are correlated with time spent on Facebook, then posts that garner a lot of "likes" imply superior outcomes. Hence, in order to optimize performance, the social media site automatically matches posts with the largest number of "likes" to the top of your content list. A Facebook user in Sri Lanka posted a video that suggested that too many shops in the town of Digana were owned by Muslims and that Singhalese should retake the town. Despite warnings by researchers in Colombo asking Facebook to take down this video, all posts remained online. "Over the next three days, mobs descended on several towns, burning mosques, Muslim-owned shops and homes." After this scene, the government blocked social media and Facebook closed the instigator's page. However, blocking was not successful. Users could access the site using virtual private network (VPN) which connects to the Internet outside the country (https://www.nytimes.com/2018/04/21/world/asia/facebook-sri-lanka-riots.html?rref=collection%2Fcolumn%2Fthe-interpreter&action=click&contentCollection=world&region=stream&module=stream_unit&version=latest&contentPlacement=2&pgtype=collection).

The above example is an illustration of supplier response in communication markets, a topic we examine in detail in Chap. 5. In the next chapter we address the demand for sharing.

References

Akerlof, David. 2016. "We Thinking" and Its Consequences, *American Economic Review Papers and Proceedings* 106 (5).

Akerlof, George and R. Kranton. 2000. Economics and Identity. *Quarterly Journal of Economics*, August, 105 (3).

Akerlof, George. and R. Kranton. 2005. Identity and the Economics of Organizations. *Journal of Economic Perspectives* 19 (1).

Carpenter, Sade. 2018. Dating While Black: Queer Black Chicagoans talk about isolation, lack of safe spaces where they can explore identity, *The Chicago Tribune*. March 15. Accessed 8/7/18 from http://www.chicagotribune.com/lifestyles/ct-life-dating-black-and-queer-20180217-story.html

Confessore, Nicholas et al., 2018. *The Follower Factory.* Accessed 6/14/2018 from https://www.nytimes.com/interactive/2018/01/27/technology/social-media-bots.html

Côté IM and Darling ES. 2018. *Scientists on Twitter: Preaching to the choir or singing from the rooftops? FACETS* 3: 682–694. doi: https://doi.org/10.1139/facets-2018-0002. Accessed 2/28/2019 from http://www.facetsjournal.com/doi/pdf/10.1139/facets-2018-0002

Damasio, Tony. 2016. *Descartes' Error, Descartes' Error: Emotion, Reason, and the Human Brain*, Kindle Edition.

Fisher, Max and Amanda Taub. 2018. This Liberal, idealist notion of the public sphere doesn't match reality. *The New York Times*. Accessed 8/28/2018 from https://static.nytimes.com/email-content/INT_1493.html?nlid=20718461

Goldberg, A., S. Srivastava, V. Govind Manian, W. Monroe, and C. Potts. 2016. Fitting In or Standing Out? The Tradeoffs of Structural and Cultural Embeddedness, *American Sociological Review*, December, 81 (6).

Gomila, R., Shepherd, H.S., & Paluck, E.L. (in prep). *Network insiders and outsiders: Who can identify influential people?*

Harare, Yuval Noah. 2017. *Sapiens: A Brief History of Humankind* and *Homo Deus: A Brief History of Tomorrow*, New York: Harper Collins.

Huettel, Scott and Rachel Kranton. 2012. Identity economics and the brain: uncovering the mechanisms of social conflict. *Philosophical Transactions of the*

Royal Society B 367. Accessed 4/29/2019 from https://royalsocietypublishing. org/doi/pdf/10.1098/rstb.2011.0264

IDEAS. Accessed on 2/27/2019 from https://ideas.repec.org/top/top.person. twitter.html

Kranton, Rachel. 2016. Identity Economics 2016: Where Do Social Distinctions and Norms Come From? *American Economic Review: Papers and Proceedings* 106 (5).

Maheshwari, Sapna. 3/1/2019. Who are Online, Recruited by Advertisers and 4-years old? Kidfluencers. *The New York Times.* Accessed 3/2/19 from https:// www.nytimes.com/2019/03/01/business/media/social-media-influenc- ers-kids.html

McClusky, Mark. n.d. *Techies are Trying to Turn the NBA into the World's Biggest Sports League.* Accessed 6/14/2018 from https://www.wired.com/2016/05/ how-tech-took-over-the-nba/

Phan, Michelle. 2007. Natural Looking Makeup Tutorial. https://www.youtube. com/watch?v=OB8nfJCOIeE

Securities and Exchange Commission Press Release. 2018. https://www.sec.gov/ news/press-release/2018-226

Tamir, Diana and Jason Mitchell. 2012. *Disclosing information about the self is intrinsically rewarding.* Proceedings of the National Academy of Sciences. Accessed 8/9/2018 from http://www.pnas.org/lookup/suppl/. https://doi. org/10.1073/pnas.1202129109/-/DCSupplemental

Taub, Amande and Max Fisher. 2018. *Where Countries are Tinderboxes and Facebook is a Match.* Accessed on 7/14/2019 from https://www.nytimes. com/2018/04/21/world/asia/facebook-sri-lanka-riots.html?rref=collection% 2Fcolumn%2Fthe-interpreter&action=click&contentCollection=world®ion =stream&module=stream_unit&version=latest&contentPlacement=2&pgty pe=collection

4

The Economics of Connectivity: Communication Markets

Standardization of connectivity is accomplished by assigning a numerical measure, follower count or likes or retweets, thereby facilitating the market exchange of a tradable product—content. Naturally, there must be buyers and sellers (see Fig. 1.3). If the pursuit of self-esteem drives most people to supply content by sharing text, images and videos, who does the actual consumption of this content? In the market for digital content the currency is time and attention. Buyers buy content and pay with time, which, importantly, is bundled with personal data. Hours of attention is the payment for consumption of content. But when every hour of attention is stamped with personally identifying information, the true payment is a composite of time and data.

The powerful confluence of streaming technology and mobile devices has made access frictionless by decoupling time and content, so control over when and where to consume is in the hands of the buyer. Mobile devices can range from the smallest smartphone to the magazine-sized lightweight laptop, both equipped with massive processing capacity. When attached to headphones, the user is subliminally ensconced in another world, effectively transforming consumption into a solitary activity. Whereas traditional media dictated both the format and the timing of a television program, content retrieval on Netflix is at our discretion.

© The Author(s) 2019
S. Bhatt, *The Attention Deficit*, https://doi.org/10.1007/978-3-030-21848-5_4

Consequently, we are both connected and alone, a paradox of connectivity—an outcome of commodifying connections and the creation of markets for communication or social media platforms. This chapter focuses on demand for content, while Chap. 5 considers the supply, provided by the media and entertainment industry.

The Advertising Model

In the early years, user-friendly implementation of digital communication technology and a cognitively friendly format on social media platforms drove demand. Subsequently, ease of use was reinforced when platforms *sold* free content. To repeat, platforms are selling *free* content.

According to a Pew Research Survey conducted in August 2018, 43 percent of US adults get their news from Facebook, 21 percent from YouTube and 12 percent from Twitter. Despite being cautious about news on social media, with 57 percent of those surveyed believing social media news to be inaccurate, 20 percent often get news from social media and another 27 percent sometimes receive their news from these sites. So 47 percent of US adults get their news from non-primary sources (Matsa and Shearer 2018). The reasons cited most often are convenience and ease of use.

The advertising-based business model supports content provision when firms pay for the advertisements that sustain the platforms. In this model, social media platforms perform a "money laundering service" by selling you content in exchange for a bundle of time and data, which is then unpacked and the data sold to advertisers for real dollars.

Content providers are independent contractors, and they need not be directly supported by the platform. These are professionals whose remuneration is job-based, measured in terms of dollars paid per hour of work on the job or a fixed price for the completed job. They are not employees as conventionally defined, since they are neither paid a wage or salary and have "an explicit or implicit contract for a continuing relationship" nor have a "predictable work schedule or predictable earnings" (Abrahamson et al. 2018).

Buyers have effortless access, at zero cost, to content laced with advertisements. There is no guarantee of quality, no overall curation, so free information could quite possibly be noisy information, including irrelevant information. Filtering out the noise from the truth requires mental effort. Effortful sorting depends upon more than raw intelligence, as shown by a recent Stanford study that found undergraduates at Stanford, among the most competitive of institutions, scored lower than professional fact checkers in simple tests, gauging the reliability of articles about bullying on websites of the American Academy of Pediatrics and the American College of Pediatricians. The former, established in 1932, is the largest professional group of pediatricians worldwide, with a membership of over 64,000. The latter is a splinter group, with 200–500 members, that broke from the parent over the issue of gay marriage and adoption. Students incorrectly, but "overwhelmingly judged the College's site the more reliable" (Wineburg and McGrew 2017).

The longer a consumer rests on a website, filtering content or generally perusing the site, the greater the revelation of personally identifying traits and characteristics. Consumers who linger on platforms are engaging with content—this is how they consume content. The value of engagement on a platform is higher when a larger group of people visit the platform, posting information and consuming information. This increase in overall platform value when more people join the network is an *externality of their joining*, an unintentional benefit to all users conferred by new users, who join simply to capture individual value.

The increase in platform value, the unintentional benefit, arises due to *network effects*, which play a vital role in the business model. Network effects arise when the services provided by the platform are more meaningful when the platform is larger. In other words, the willingness to pay for a product or services provided by a platform increases with the number of people consuming this service. As the subscriber count for YouTube rises, YouTube's value to current users increases since each subscriber can now engage with a larger audience. New subscribers to YouTube are attracted by the benefit of social engagement, and they unintentionally confer the externality of a larger subscriber base upon joining. It follows that the more users there are on the social media platforms, the greater the reach of advertisements, incentivizing marketing firms to shower more revenue onto the media platform.[1]

In order to increase time spent on their site, platforms employ a variety of subtle and not-so-subtle strategies. If a consumer lingers on the site of the American College of Pediatricians, for example, checking out recent articles on bullying, it is likely that messages from Planned Parenthood and similar sites will then be directed toward this consumer. The longer a consumer lingers on a site, the higher the likelihood that this user will click on an advertisement and even make a purchase. The organization posting the ad incurs a cost-per-click when a customer clicks on the link and accesses the site. Further payments arise upon actual transactions on the site. Consequently, the platform has an incentive to increase your overall time spent on their site since that increases advertising revenue.

Snapchat, the social networking platform, has a number called "streaks" which measures how many continuous days you have connected with a friend. The longer the streak, the stronger the friendship. Frequency of communication is a measure of friendship. Even more important for the platform is the fact that longer streaks imply a larger total amount of time spent on their site. The depth of that relationship, measured in terms of shared experiences, is irrelevant to this calculation. Suppose I have a year-long streak with person B, but I spend the entire summer hiking the Rockies with person A, and then not reconnect with A on Snapchat until the following summer. Does person B rank as a good friend or just a passing acquaintance?

The advertising business model has been studied by Tim Wu, who illustrates the role of the penny press in democratizing news. Benjamin Day started the *New York Sun* in the 1830s by charging pennies for the publication, amassing a large readership, which was then used to attract the true revenue source—advertisers (see Chap. 5 for more detail). Clearly the larger the audience for content, the more valuable it is to advertisers, so the goal is to make content seductive:

---

[1] Technically, these are indirect network effects, since advertisers are providing a complementary product, advertisements, whose reach, and therefore payoff, increases with market size. Advertisers are more incentivized to pay for ads on larger platforms than smaller platforms. Consequently, buyers on larger platforms gain, indirectly. Direct network effects arise when the value of the original product increases with the market size, as in telephone networks. More details follow in Chap. 5. Also see Bhatt (2017, p. 50).

This means that under competition, the race will naturally run to the bottom; attention will almost invariably gravitate to the more garish, lurid, outrageous alternative, whatever stimulus may more likely engage what cognitive scientists call our "automatic" attention as opposed to our "controlled" attention, the kind we direct with intent.

However, at some point there is the "disenchantment effect … [which] happens when a once entrancing means of harvesting attention starts to lose its charm" (Wu 2016).

Competition for resources has shifted from the media product market (where the product is you) to the financiers of this market—the advertisers. What you consume, how you consume and when you consume is decided upon the convenience of the sponsors. Digital advertising (both desktop and mobile) was 44 percent of all advertising revenue in 2017 compared with 37 percent in 2016; mobile digital advertising had twice the budget as desktop digital—$60.7 billion compared with $29.7 billion for desktop (Pew Research Center 2018).

Ads that are displayed alongside content are digital display ads (banner ads, rich media ads or ads with audio/video components) and they dominate all forms of digital advertising. Of the $60.7 billion spent on mobile digital advertising, $31.6 billion was allocated to mobile display advertising. Facebook dominated this digital display advertising segment with 39 percent with Google trailing at 13 percent; Facebook also captured half of the *mobile* display ad segment (Pew Research Center 2018).

The digital advertising market is becoming more of a monopsony as media behemoths have superior buying power, while other companies are downsizing. BuzzFeed recently announced that it was laying off 15 percent of its workforce, or 250 jobs. Verizon is laying off 7 percent of staff, or 800 jobs, from the Verizon Media Group, which includes AOL and Yahoo (Winkielman and Cacioppo 2001). Both are facing stiff competition for advertising dollars. BuzzFeed's founder, Jonah Peretti, a graduate of Massachusetts Institute of Technology (MIT) Media Lab, had a goal of creating shareable news on social media when he created the firm in 2006. But Facebook retaliated by reducing the prominence of professional news sources such as BuzzFeed, Vox Media and Refinery 29 (Lee 2019).

The Psychologists' Mental Accounting Model

Most people perform a sort of mental accounting for time, which provides a framework for thinking about the true cost of content in the advertising model. Daniel Kahneman's Nobel Prize-winning work on mental heuristics, posits a system 1 and system 2 level of cognitive effort. System 1 requires minimal mental effort, while system 2 "allocates attention to the effortful mental activities that demand it, including complex computations" (Kahneman 2011). Adapting and generalizing his model, I postulate two mental accounts in the time vault: h_a, the rationally chosen hours of attention and h_l, the rationally chosen hours of leisure, with $h_a + h_l = 24$. While casual perusal of information may draw on the leisure account, focus and processing of information draws on the attention account.

Social and commercial media provide digital content that captures buyers in a distinctive way. The content is easy to process and utilizing minimal mental effort may well dip into the leisure account. Cognitively friendly formats make "thinking easy" or engender fluency in information processing, so social media generates positive affect or a general feeling of well-being. If information is easy to process, then comprehension of the idea is achieved faster, eliciting the rewarding feeling of achievement. Moreover, intuition, working like an "associative machine" that represents reality by a simple chain of links, equates familiarity and repetition with cause-and-effect patterns (Kahneman 2011). Recommendation algorithms are a case in point. *Recommended for you* or *people who bought x also bought y* are highly suggestive phrases. There is the familiarity effect when grouped with similar people and there is a cause-and-effect when similar individuals made follow-up choices that are being recommended. Promotional content, converted into fluid, engaging material, captures the mind and impacts brand loyalty by building long-term brand equity. It increases revenue across all of the firm's product categories, suggesting that social media engagement strengthens buyers' relationships with the firm (Kumar et al. 2016). This could lead to unfortunate and unreasonable conclusions, such as overconfidence and manipulability by carefully framed arguments, when the situation requires thoughtful analysis based on available information and statistical reasoning.[2]

---

[2] "Research in cognitive psychology suggests that repeated exposure facilitates stimulus [information] processing. This is reflected in faster stimulus recognition, higher judgements of stimulus

When the content contains noisy information, users require mental effort to extract and absorb the "true" content. Mental effort is required to sort out useful content and ignore useless content, so information overload is an acute problem for people with limited curating power. Exercising self-control to avoid going down rabbit holes of wasteful content itself consumes attention; effortful sorting is imperative for filtering out superfluous information. The ability "to block out most everything and focus— is what neuroscientists and psychologists refer to as paying attention." Moreover, "[D]epending on the kind of information, it takes our brains some amount of time to process it, and when we are presented with too much at once we begin to panic" (Winkielman and Cacioppo 2001).

In terms of mental accounting, the activity of sorting and filtering requires effort and so available hours of attention are less than rationally chosen hours. Activity on social media requires constant monitoring. Pleasurable feelings of well-being or positive self-esteem morph into a craving or addiction.[3] This requires self-control or actively terminating connectivity, by shutting down our devices or logging off. But self-control itself requires mental effort. According to Kaushik Basu, individuals may be rational in their management of time, but they have vulnerabilities and "impulses that are hard to control."

> It is as if human beings had monkeys on their shoulders who make the choices for them, but not always in their interest. This creates an opportunity for the scheming and the ruthless to take advantage of consumers by enticing, misleading and misinforming them. (Basu 2018)

Since total available hours are a true hard constraint, an *attention scarcity* is created, a deficit of *cognitive bandwidth hours or attention*. There follows a collective cognitive breakdown, an inability to selectively absorb relevant information.

---

clarity and duration." Further evidence is provided in studies that observed participants evaluate pictures of objects and found that "the participants indicated higher preference for easy-to-process pictures." This suggests that individuals prefer to consume information in easy-to-understand sound bites (Winkielman and Cacioppo 2001).

[3] Addiction to social media arises when the liking is transformed into wanting, when there is a desire to extend the pleasurable sensation of dopamine in the brain synapses for a longer period.

The Economists' Model

What is the true price of social media consumption? Economists model the allocation of waking hours between work and leisure in a labor-leisure choice model, analogous to the mental accounting of time described earlier. The price of an hour of leisure is the hourly wage rate since you give up an hour's worth of wages to buy an *extra* hour of leisure.

Mas et al. estimate that individuals value their leisure hours at 60 percent of their gross wage, but that the value of marginal leisure time varies by hours of work. "The lowest wage required for a worker to accept a 20-hour job will be lower than the lowest wage required for a worker to accept a 40-hour job." This suggests that an individual who works 20 hours a week, values leisure hours less than another individual who works 40 hours a week, since the former has more leisure to begin with. At higher incomes, individuals value their leisure at a higher rate (Mas and Pallais 2017a).

Flexible work arrangements may have actually increased leisure hours by enabling a reallocation of work hours. For example, by working from home or commuting during non-rush hours, people can save on travel time. However, in another study, Mas finds that flexibility in work hours is not highly valued by the majority of workers, but that "workers do value the option to work from home and strongly dislike employers setting their schedules on short notice, mainly because they don't want to work evening and weekends. Overall, the traditional M–F 9am–5pm schedule works well for most people, perhaps because this schedule allows them to coordinate their leisure time" with friends as well as with media schedules (Mas and Pallais 2017b).

Choices among competing demands are made by comparing relative benefits against relative costs. How do we choose hours of social media consumption? We have seen the benefits of social media consumption, but what is the cost of an extra hour of social media perusal? It is the cost of producing that hour of social media perusal: the time cost (or the other opportunities forfeited for that activity). We tend to focus on benefits, ignoring the costs. If you give up one hour of coffee time with an old friend for an hour with YouTube, then the price of that hour of social media consumption is the value you place on conversation with that friend. Most of us have not placed an exact value on this friendship, so we are unable to make rational choices. We may underestimate the value of friendship and hence the true price of social media consumption.

Frequently, the time allocated to social media is not the result of conscious, rational allocation as described earlier. What about services that automatically do this allocation for us by usurping our attention for content we did not choose? This can confound your time allocation effort, as when your YouTube time is no longer under your control. For example, suppose you were a writer intensely working on a new novel and lost track of time as you worked beyond the original four hours you had allocated. You forfeited dinner, a walk or conversation with a friend. Is this any different from YouTube overtime? Yes. As a writer, the monetary or other rewards from an extra hour of working on the novel are positive and possibly exceed the emotional rewards of conversation. On the other hand, when I am subtly drawn into YouTube overtime, the choice is made for me with uncertain rewards.

Artificial intelligence has increased leisure hours by efficiency enhancements impounded in everyday devices.[4] Wearable devices have become a personal concierge, providing guidance for daily functioning. Careful calibration of hours of sleep, intake of calories, steps walked, heart rate and blood pressure monitor life styles automatically, requiring little mental effort from the individual. Further, time-saving devices endowed with recommendation algorithms reduce cognitive effort and provide more leisure time. Personalized recommendations for purchases on Amazon and Netflix, and even rides for Uber (based on prior history), for example, provide faster, more efficient customer satisfaction.

Individuals have more leisure as devices automate, synchronize and anticipate time-consuming chores. Since "one of AI's main effects will be a dramatic drop in the cost of making predictions," product recommendations and targeted advertising make choices less time consuming (Economist 2018). Efficiency in digital payments has increased due to

---

[4] I use the word artificial intelligence to refer to the science or theory of making intelligent machines and machine learning to denote the algorithms and computer systems that self-learn without being explicitly programmed. Algorithms derive inferences and make predictions based on pattern recognition, speech recognition and facial recognition. Importantly, these algorithms exhibit emergent behavior by learning from new data.

advances in computer vision and speech recognition, streamlining identification and verification. Apple Pay's thumbprint signature is going to replace handwritten signatures. Paradoxically, the very same platforms (Amazon's Alexa and Google's Home) that sell you these time-saving devices also give you the currency, your time-use profile, to pay for their content. Every move you make is noted and recorded by the platform!

The evidence supports the notion that leisure hours have increased in recent years and that more leisure is allocated to entertainment. According to the data from the American Time Use Survey, total non-work hours, devoted to leisure and sports, including travel, have increased in 2017.

Figure 4.1 shows *overall leisure hours per day*, week-day plus weekend, from 2003 through 2017. The peak is in 2012 at 5.37 hours, with a downward trend that is dramatically reversed in 2017, going from 5.13 hours in 2016 to 5.24 hours in 2017.

Figure 4.2 shows *weekend hours* per day devoted to leisure and sports, including travel, from 2007 to 2017. The peak is over 6.57 leisure hours per weekend day in 2012. There is a downward trend since 2012, with a reversal in 2017, going from 6.43 hours in both 2015 and 2016 to 6.46 hours in 2017.

Not only have leisure hours increased, but consumption of leisure time services has increased. Consider Fig. 4.3, which describes consumer expenditure by age group, on entertainment, health care and education, for the year 2016. The 35–64 age demographic, the boomers, spent most (greater than $3000 annually) on entertainment broadly defined. Figure 4.4 depicts changes in annual expenditures over the period 2013–2016. Consumer expenditure for all age groups has increased in all categories. The numbers are not inflation adjusted. Remarkably, expenditures on entertainment and education have grown at roughly similar rates over the 2013–2016 time period, suggesting the important role of the entertainment industry. Expenditure on entertainment has increased by 17 percent, going from $2482 per year in 2013 to $2913 in 2016. Education expenditure has grown by 16 percent, rising from $1138 to $1329 per year. As expected, health-care

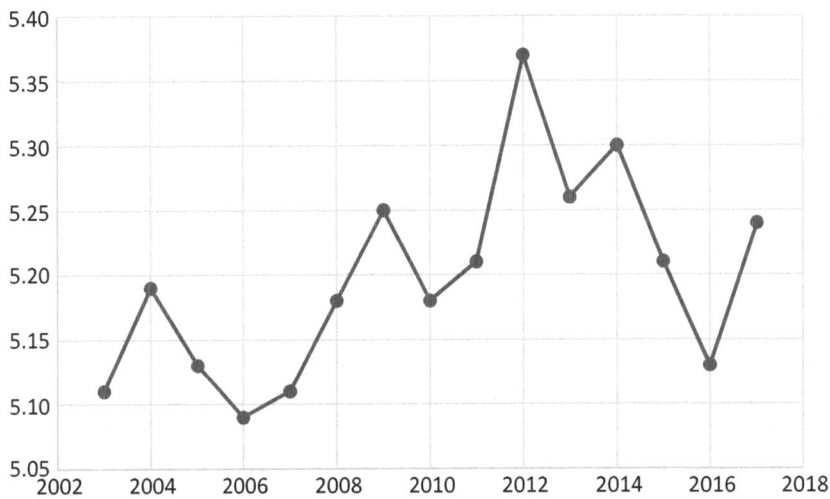

Fig. 4.1 Average hours per day—leisure and sports (includes travel). (Source: Bureau of Labor Statistics, American Time Use Survey. Retrieved on July 16, 2018, from https://www.bls.gov/tus/)

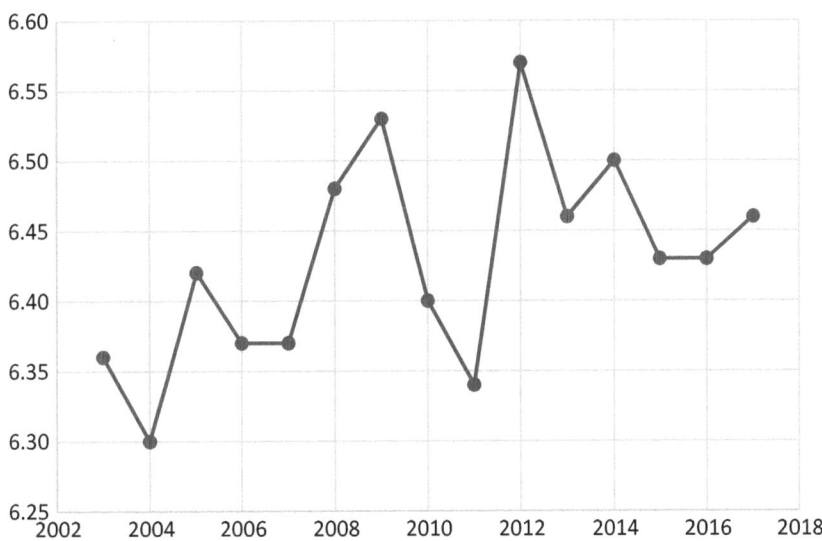

Fig. 4.2 Average hours per day—leisure and sports (includes travel, weekend days and holidays). (Note: Leisure and Sports = (Socializing and Communicating) + (Watching Television) + (Participating in sports, exercise, and recreation). Source: Bureau of Labor Statistics, American Time Use Survey. Retrieved on July 16, 2018, from https://www.bls.gov/tus/)

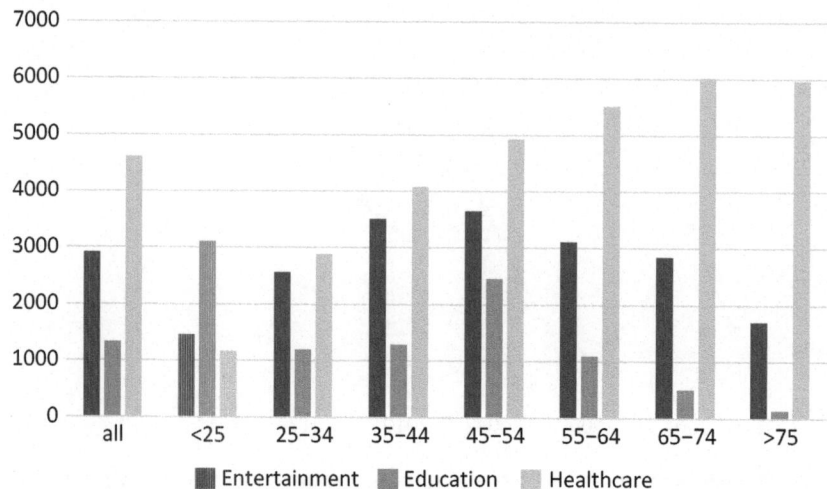

Fig. 4.3 Annual consumer expenditure in 2016. (Source: https://www.bls.gov/cex/2016/combined/age.pdf. Retrieved 8/7/2018)

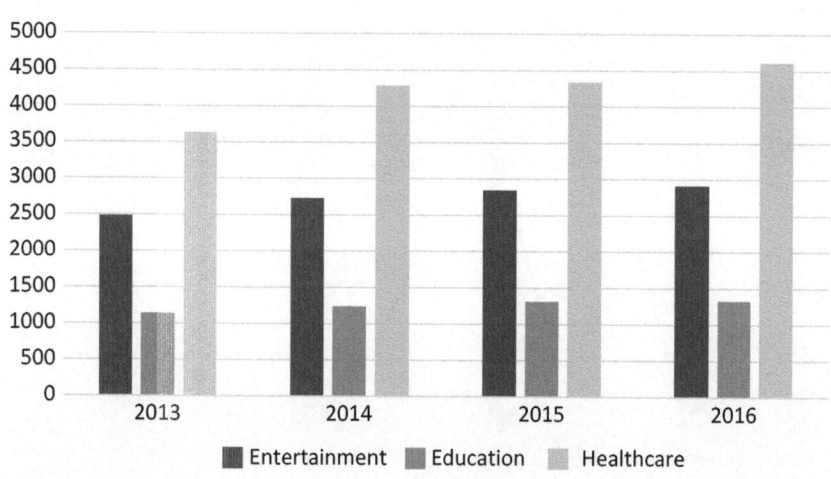

Fig. 4.4 Annual consumer expenditures 2013–2016. (Source: https://www.bls.gov/cex/2016/combined/age.pdf. Retrieved 8/7/2018)

expenditures have increased dramatically going from $3631 in 2012 to $4612 in 2016, an increase of 27 percent.

The entertainment industry, well aware of this increase in leisure, has leveraged streaming technology, along with software for creations in virtual reality and augmented reality, and blasted the world with content.

References

Abrahamson, Katherine, John Haltiwanger, Kristin Sandusky and James Spletzer, December 2018. *The Rise of the Gig Economy: Fact or Fiction.* US Census Bureau Working Paper.

Basu, Kaushik. 2018. Markets and Manipulation: Time for a Paradigm Shift, *JEL* 56 (1), March.

Bhatt, Swati. 2017. *How Digital Communication Technology Shapes Markets: Redefining Competition, Building Cooperation.* New York: Palgrave Macmillan.

Economist. 2018. *Non-Tech businesses are beginning to use artificial intelligence at scale.* Retrieved from https://www.economist.com/news/special-report/21739431-artificial-intelligence-spreading-beyond-technology-sector-big-consequences

Kahneman, Daniel. 2011. *Thinking Fast and Slow*, New York: Farrar, Strauss & Giroux, 2011.

Kumar, Ashish, Ram Bezawada, Rishika Rishika, Ramkumar Janakiraman and P.K. Kannan. 2016. From Social to Sale: The Effects of Firm Generated Content in Social Media on Customer Behavior. *Journal of Marketing* 80, January.

Lee, Edmund. January 23, 2019. BuzzFeed Plans Layoffs as It Aims to Turn Profit. *New York Times*, Accessed 1/24/19 from https://www.nytimes.com/2019/01/23/business/media/buzzfeed-layoffs.html

Mas, Alexandre and Amanda Pallais. 2017a. *Labor Supply and the Value of Non-Work Time: Experimental Estimates from the Field,* Working Paper, Princeton University, September.

Mas, Alexandre and Amanda Pallais. 2017b. *Valuing Alternative Work Arrangements,* Working Paper, Princeton University, March.

Matsa, Katerina and Elisa Shearer. 2018. *News Use Across Social Media Platforms 2018,* Pew Research Center. Accessed 1/24/19 from http://www.journalism.org/2018/09/10/news-use-across-social-media-platforms-2018/

Pew Research Center. 2018. Digital News Fact Sheet. Accessed 1/24/2019 from http://www.journalism.org/fact-sheet/digital-news/

Wineburg, Sam and McGrew, Sarah. 2017. *Lateral Reading: Reading Less and Learning More When Evaluating Digital Information* (October 6). Stanford History Education Group Working Paper No. 2017-A1. Available at SSRN: https://ssrn.com/abstract=3048994 or https://doi.org/10.2139/ssrn.3048994

Winkielman, Piotr and John Cacioppo. 2001. Mind at Ease Puts a Smile on the Face: Psychophysiological Evidence That Processing Facilitation Elicits Positive Affect, *Journal of Personality and Social Psychology* 81, pp. 289–1000.

Wu, Tim. 2016. *The Attention Merchants*, New York: Alfred A. Knopf.

5

Streaming Technology and the Entertainment Industry

Sellers of digital content, likely anticipating the increase in leisure and entertainment demand, have increased digital content, unleashing a content tsunami by providing more services (information, media and entertainment) and more ways to consume these services. ICT has changed the nature of the entertainment industry due to three developments: streaming technology, indirect network effects and low production and distribution costs of user-generated content.

Digitization has relaxed the constraint on content distribution. Digital subscription and streaming music revenues accounted for 79 percent of the global recorded music market in 2017 (RIAA 2017a). Pop music is the most popular music genre in the world, with rock and dance/electronic music as the second and third most popular, and hip-hop and rap as fifth (Statista 2019). The most streamed music across the globe, by country of origin, is from American artists. Of the top ten global recording artists in 2018, all except three (BTS who are from South Korea, and Ed Sheeran and Queen, who are British) are US citizens and Drake, who has dual US-Canadian citizenship, has been named the world's best-selling recording artist for 2018 for his album *Scorpion* (IFPI 2019). On the video side, US-origin film repertoire has the highest global penetration both in theaters and on global platforms such as Netflix (Aguiar and Waldfogel 2017).

© The Author(s) 2019
S. Bhatt, *The Attention Deficit*, https://doi.org/10.1007/978-3-030-21848-5_5

Streaming

The grand decoupling of time and content via streaming technology and mobile devices has been transformative. Decoupling means the time delay between the final consumption of content and the initial urge to view content is controlled, to a large extent, by the user. Prior to the streaming era, time was matched to content by the platform and there arose a gap between actual consumption of content and the moment of desire to consume content, a time delay. Today, the platform no longer determines when various shows or videos can be watched nor when songs can be heard.

In principle, all music and all movies can be made available everywhere. In 2013, cementing the decoupling of time and content and demise of linear television by streaming, Netflix commissioned and uploaded 26 episodes of *House of Cards*, 13 per season all at once. It went on to run for six seasons from 2013 to 2018. Control over content was now in the hands of the viewers; when, where and what to watch was no longer determined by broadcast television.

By making time delay a user-determined meter, streaming technology has conferred control over time allocated to content consumption to viewers. Reallocation of control has restored agency in entertainment choices and increased video viewership since users can now follow the story line uninterrupted. This shift in demand has precipitated two major changes: (1) a shrinking interval between theatrical release dates and streaming release dates and (2) binge-watching due to collapsed production schedules. Both generate the capability of solo, anytime and anywhere consumption as consumers move away from communal entertainment to solitary consumption. The FOMO factor adds to the demand for binging, since consumers desire to keep abreast of new jargon. Media platforms perform distribution services alone rather than control the timing of distribution and dissemination. This trend has shifted power from the platform to the user and content creator. Some platforms, such as Netflix, are flexing their power in response to this development, by vertically integrating into content creation.

Setting aside piracy (which is a major issue for artists), content can be accessed in two ways: rent via subscription (Netflix, Amazon Prime) or purchase via a pay-as-you-go a la carte model (Amazon Instant, Apple). The subscription model has a flat fee for consumers for unlimited access and a flat fee for content producers for a given catalog, chosen by the platform. In the a la carte model, platforms share revenue with producers by paying a 70 percent royalty fee per piece distributed, so the catalog is wider and curation by platform is limited. Consumption of content, by contrast, is larger for subscription services since the price of an additional firm is zero.

Streaming originated with video and audio on demand, where the user chose the time, rather than suffering through the time delay when broadcasters determined the viewing at a predetermined time. Content was now referred to as streaming video-on-demand or SVOD and streaming audio-on-demand or SAOD. Moreover, streaming technology enabled users to view video or hear audio without downloading the entire digital video or audio file, so there was fee-based access without the costly burden of ownership. Live streaming is delivery of content in real time over the Internet, much like a live broadcast over television. Since its introduction in 2016 on Snap's Live Stories, on Facebook Live, on Twitter's Moments and Periscope and Google's YouTube, people brought short, personalized videos to mobile devices.

RealNetworks, based in Seattle, was one of the early companies that created streaming technology, followed by Microsoft's Window's Media Player in 1999 and Apple's QuickTime 4 application. The widespread use of streaming technology, however, awaited the arrival of Netflix, Spotify, Pandora and Amazon Music in early 2006. Together, these developments have made it easier for small independent producers to gain market share. The jury is still out on the question of whether these developments have increased innovative content or merely enable efficient recycling of stale content. Of note is the fact that the Amazon-funded Moonlight, with a production budget of under $1 million, won 2017 Oscar for Best Picture. Netflix's Roma won the 2019 Oscar for Best Director and Best Foreign Film. Both firms trace their origins as OTT or over-the-top (of the existing Internet service providers' infrastructure) distribution firms but have now integrated backward into content production.

Network Effects

When more users engage on the platform, more content creators are incentivized to provide content and more advertisers are motivated to insert ads since there is a wider audience. These are *cross-side* *indirect* *network effects* because larger content catalogs encourage more users to engage by increasing value of the platform. The sellers, using an advertising-based model, provide free content.[1] While the advertising-based media model predates the digital revolution, ICT facilitates real-time advertising such that a given user's current content is perfectly aligned with the advertising message sent. If you are searching for shoes, you will most likely be inundated with ads from shoe manufacturers.

Purveyors of content harness the power of network effects to build their library. Media platforms have multiple constituencies, when they interface with consumers on one side, with content creators on a second side and advertisers on the third side. In multi-sided platforms, to generate value, all sides need to keep growing—buyers need to grow to feed the advertisers and content creators, and advertisers, in turn, support creators. The platform's goal is to increase demand and generate revenue. Cross-side network effects refer to the spillover or externalities generated by the actions of one party on the welfare of other parties. The platform is essentially selling virtual real estate to two sides: content creators and advertisers. But only the latter party pays.[2]

Low Production and Distribution Costs

To meet this growth in content, the entertainment industry is on a hiring and merger binge. The overall employment growth rate is projected to be 7.4 percent over 2016–2026 (Bureau of Labor Statistics 2017).

---

[1] Premium subscriptions to digital content generate revenue by separating out buyers who are willing to pay a high price to avoid advertisements; the remaining buyers are offered the free version with ads.

[2] Zeynep Tuftci writes, "These companies … use massive surveillance of our behavior, online and off, to generate increasingly accurate, automated predictions of what advertisements we are most susceptible to and what content will keep us clicking, tapping, and scrolling down a bottomless feed" (Tuftci, 2019).

However, for the entertainment industry, the growth rate is higher. The Employment Projections program at the Bureau of Labor Statistics calculates projected average employment as 7.6 percent over 2016–2026, for more than 800 video-related occupations, where video encompasses the production, recording and rebroadcast of visual images via theaters, traditional broadcast (including cable and satellite) television and streaming services over the Internet. In particular, film and video editors will see the largest growth rate at 17 percent over this time period, while actors have a 11.6 percent employment growth rate and producers and directors will see a 12.2 percent rate. Streaming content over mobile devices is the primary driver of this surge in video consumption (Bureau of Labor Statistics 2018).

Churning of entertainment industry assets supports this notion of accelerating growth of the industry. While Amazon's $13 billion acquisition of Whole Foods appears to be a purchase of customer data, AT&T's merger with Time Warner and Disney's acquisition of 21st Century Fox assets have analysts betting on Apple purchasing a major studio, such as J.J. Abrams' Bad Robot Studio. Hulu, which is 60 percent owned by Disney Fox, with additional ownership stakes by Comcast (30 percent through NBC Universal subsidiary and 10 percent by Time Warner), could be in play if either of the latter two took charge. Hulu-NBC Universal could be a serious challenger to Disney, Netflix and Amazon in the streaming video industry (Gomes-Casseres 2018). We discuss mergers and acquisitions in the motion picture industry later.

Low production and distribution costs on YouTube and SoundCloud, for example, have enabled self-publication across multiple media formats, such as print, audio, video and online text. YouTube encourages the budding film director but also indiscriminately lets content explode onto users' screens. ReverbNation helps discover new musicians with a data management tool, Band Equity, that ranks new artists on four dimensions: reach, impact, ease of access and release date. The Gig Finder tool provides a listing of potential venues for new artists.

The Evidence

Separating the video and audio market, let us examine the video industry, motion pictures and video games, first.

The Entertainment Industry: Motion Pictures

Data from the American Time Use Survey suggest that leisure time is increasingly dominated by screen time (see Figs. 5.1, 5.2 and 5.3, components of Fig. 4.2).

Screen time includes the following activities: television and movies, leisure computer use, email and messages, video games and in-theater movies (Krause and Sawhill 2016). Apart from the detrimental impact on health and grades, it has been shown that excess screen time promotes antisocial behaviors. The sequential nature of online activity enhances feelings of isolation because time delay allows for more metered communication and less spontaneity. When you replace face-to-face

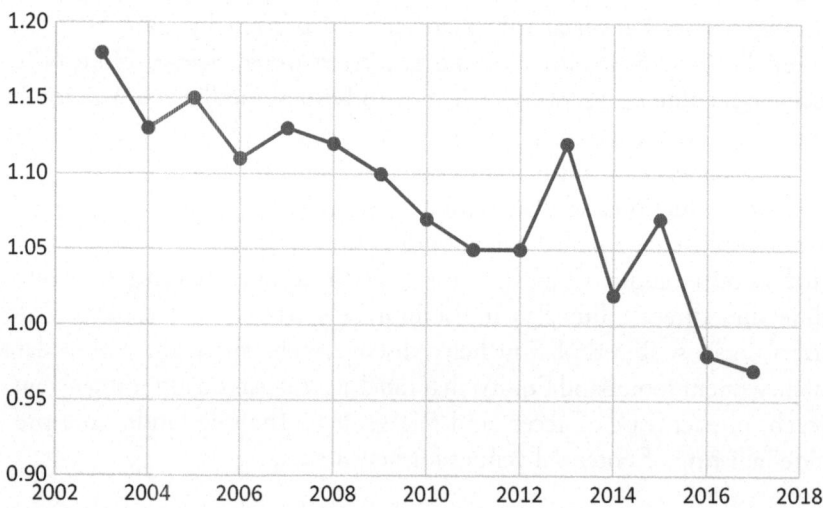

Fig. 5.1 Average hours per day—socializing and communicating. (Source: American Time Use Survey, Bureau of Labor Statistics. Retrieved on 6/5/2019 from https://www.bls.gov/tus/)

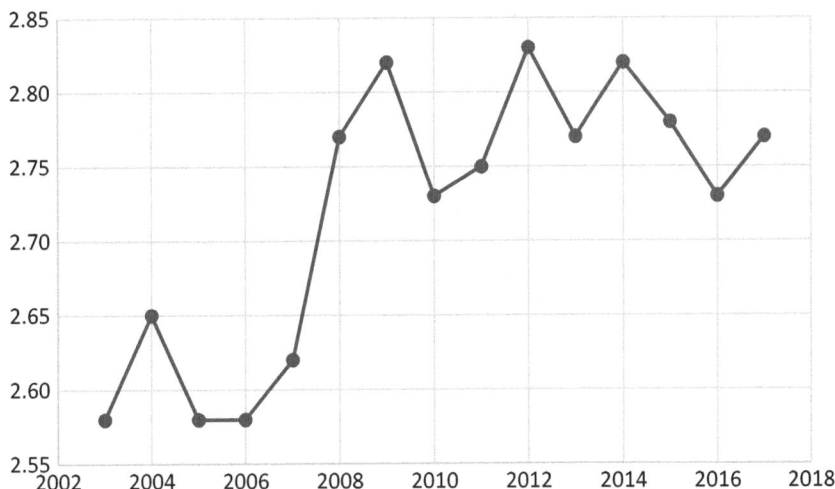

Fig. 5.2 Average hours per day—watching TV. (Source: American Time Use Survey, Bureau of Labor Statistics. Retrieved on 6/5/2019 from https://www.bls. gov/tus/)

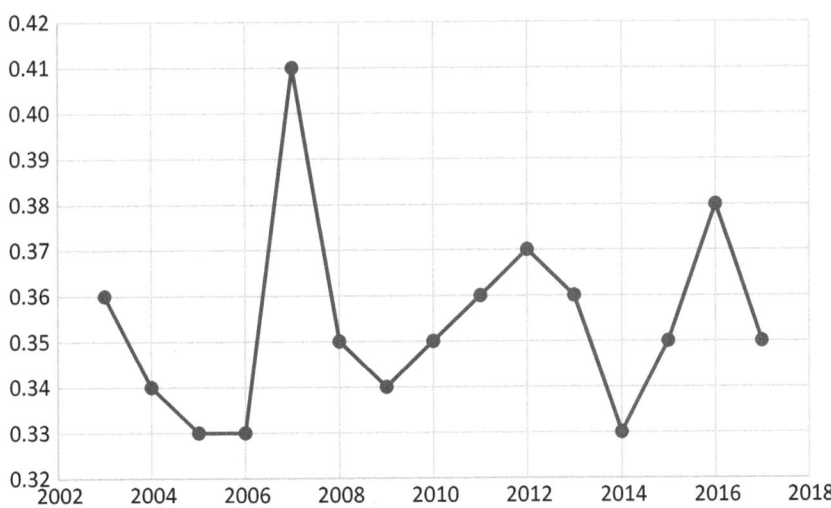

Fig. 5.3 Average hours per day—participating in sports, exercise and recreation. (Source: https://www.bls.gov/tus/. See the following for definitions: https://www. bls.gov/tus/lexiconnoex0317.pdf)

communication with screen-based communication, your reply to email or text messages allows poetic license or creative manipulation of words (Baumeister et al. n.d.).

Streaming technology and global markets have enabled an explosion in demand for moving pictures. In particular, the franchise model, with superheroes, has replaced individually created dramas and comedies. Studios have increased production budgets exponentially and are producing more sequels than purely innovative video productions. But Netflix is becoming the dominant player, with a production budget for content at $7–8 billion in 2018 (Koblin 2017).

Netflix Netflix launched its website in April 1998 based on the rental model which morphed in September 1999 into the monthly subscription model. In May 2002 Netflix was ready for public markets and initiated the initial public offering (IPO). However, Amazon introduced video-on-demand or streaming in September 2006, beating Netflix by three months—Netflix announced streaming in January 2007.[3] The year 2007 was a watershed year for entertainment with the introduction of the smartphone, the peak of digital versatile disc (DVD) sales and the brink of the Marvel cinematic universe explosion with *Iron Man*, released in 2008 (Fritz 2018).

Netflix's goal was to take streaming of art-house films out of the theaters and into homes, saving on marketing and publicity. Here again, Amazon beat Netflix in supporting indie films: its $10 million production of *Manchester by the Sea* won the 2017 Best Picture Oscar. Moreover, "Amazon found an identity with Transparent, a groundbreaking show about a transgender woman. … It also gave Amazon Studios an identity as a home for highbrow series aimed squarely at upscale, educated, affluent types" (Fritz 2018).

As a global distribution platform, Netflix showcases films across nations, acting like an intermediary. The Swiss film *Tibetan Warrior* was distributed in 243 countries, for example, and the Hong Kong film *Ip*

---

[3] The first in-house production, House of Cards, was streamed in February 2013.

Man in 103 countries. US-origin content, however, has the largest distribution through Netflix as well as the largest theatrical distribution (Aguiar and Waldfogel 2017). Aguiar and Waldfogel have found, using 2008–2014 data from IMDB (Internet Movie Database), that "US origin fare has the highest value weighted geographical reach" in theaters, followed by France, the UK, Australia and Germany. Similarly, US-origin repertoire has the highest reach on Netflix, followed by Australia, Hong Kong, Mexico and the UK—over 50 percent of theatrical distribution of films and Netflix distribution of movies plus TV series were of US origin (Aguiar and Waldfogel 2017). Theatrical reach outperforms Netflix reach in almost all countries, where reach is the proportion of the country's population that has access to value weighted content.[4]

Not only does Netflix have gatekeeping power of creative works by other studios, it has even more curating power on the creative side. As of 2019 Netflix plans to release 90 original films and documentaries a year with a budget of $200 million per film. By comparison, Universal releases 30 movies per year (Barnes 2018).

The development of a distribution platform with global penetration raises issues of cultural dominance as Netflix's signature programs are from the US and about the US. To the extent that Netflix will pursue a global strategy to produce and widely distribute US programming, "protectors of local cultural production" will likely impose trade barriers. "As the Netflix audience evolves—as the service attracts subscribers in its 243 operating countries—the composition of its audience will change. This may, in turn, affect the prospects for programming hailing from different origin countries and appealing to different kinds of viewers" (Aguiar and Waldfogel 2017).

Territoriality in the distribution of motion pictures undermines creation of a digital global market (DGM). The European Union (EU), for example, forbids iTunes France from selling to consumers in Italy and Netflix Germany from selling to users in Spain. The fear is that multinational distribution might make films targeted to narrowly defined local markets unprofitable when there is substitutability across this medium. If users in

---

[4] Global reach is calculated by summing the product of value weights for each movie multiplied by the share of the country's population that has access to that movie.

Spain are equally comfortable viewing films made by Spanish producers and those made by German producers, then a DGM induces greater competitive threats (Aguiar and Waldfogel 2017).

Marketing costs are low when data on consumer preferences allow precise targeting of movie trailers. While Netflix has shrunk the time between theatrical release and online release to zero, so that there was simultaneous release across all media, Amazon followed the traditional model of waiting for five months before transitioning to Amazon Prime. Only the truly passionate viewers would see the theatrical release despite the convenience offered by in-home streaming—a Netflix strategy.[5] However, *Roma*, winner of the 2019 Best Foreign Film Oscar, was released in theaters on November 21, 2018, fully three weeks prior to online release and Amazon's first release, *Chi-Raq*, was on the big screen for 30 days in December 2015. Steven Spielberg has expressed concern about substitution of the motion picture theatrical experience with the streaming home-viewing experience. "I don't believe films that are just given token qualifications in a couple of theaters for less than a week should qualify for the Academy Award nomination" (Barnes 2018). To what extent are the two substitutes—what qualifies as a movie and therefore can be considered for big awards. To the extent that Hollywood studio producers were concerned about Netflix's *Roma* taking a seat at the Oscar table suggests that these two are excellent substitutes so that the competitive threat can only be mitigated by staggering the availability of the two formats.

Studios Just as Hollywood was being threatened by the streaming revolution and Netflix, China came to the rescue. "China is now the wallet. And Hollywood is the factory" (Fritz 2018). As the Chinese economy shifted from manufacturing to consumption of goods and services, "Chinese audiences went crazy for Hollywood films, particularly big-budget franchise ones" (Fritz 2018). The trend started in 2002 when IMAX opened its first theater in Tiananmen Square, Beijing. After James Cameron's *Avatar* grossed $204 million in China, content was supplied with "the Chinese market in mind. Chinese audiences couldn't understand and weren't interested in most American dramas, comedies" or difficult to

---

[5] Amazon was tapping into its existing client base: "the people who go to art-house movies tend to be upscale, well-educated people who live in cities and who like to shop online" (Fritz 2018).

translate films. So branded franchise films were being produced. On the supply side, studios were being financed by Chinese investors. "They invested in Steven Spielberg's new company, Amblin Partners" (Fritz 2018) as Dalian Wanda purchased AMC theaters and the studio, Legendary Pictures. Investing in the movie business is a high-risk strategy, so what was the underlying motive? Fritz writes:

"But those who worked most closely with the Chinese knew that the biggest reason for these investments was a form of reverse-colonialism. After more than a decade as a place for Hollywood to make money, China wanted to turn the tables." Moreover, "The Beijing government considered art and culture to be a form of 'soft power,' whereby it could extend influence around the world without the use of weapons" (Fritz 2018).

Outside, financing restructured the industry by unbundling finance, production and distribution, providing a mechanism for serious downsizing of what were once giant studio-conglomerates with everything done in-house. A studio executive with a vision could now produce a mid-budget art-house movie with Chinese finance and a major studio's distribution resources. But more likely, this executive would oversee the development of the idea and then consider different platforms for the actual implementation: television or film.

Vertical integration in the film industry has brought content distributors and content creators together. Studios, broadcast television and cable television/Internet service providers are merging on a massive scale. As discussed earlier, AT&T, which merged with Direct TV in 2015, has recently merged with Time Warner for $85 billion. Disney's horizontal merger with 21st Century Fox's entertainment assets (movie studio, cable channels and stake in Sky; the sports and news divisions are to be spun off and remain with the Murdoch family) has been recently approved by the Justice Department.

The end result is a reincarnation of the classic studio system, with a class of conglomerate each controlling their own pipeline from talent to camera to screen. It's a very different vision of online video than the previous generation, where tech-focused companies could build a service based entirely on streaming tech. Now that technology is just one link in a much longer chain, and the entire industry is reshaping itself in response. (Brandom n.d.)

Of note is that Comcast, which had contested the merger, is the modern-day legacy of the golden age of movies, which lasted from 1928 to 1949, when Universal Studios was one of the big eight—Loews/MGM, Paramount, Fox, Warner, RKO, Universal, Columbia and United Artists. The interwoven nature of the industry is astonishing when one considers that Comcast is one of Disney's biggest distributors and a fellow shareholder in Hulu, an online streaming service (distributor). Hulu has old media parents: NBCUniversal, Disney/ABC, Fox, Comcast and Time Warner. This suggests that the studios are partnering with broadcast television to consolidate the creative side of business.

Are Disney, Netflix and AT&T-Warner competing for status as the new industry behemoths? Should the Justice Department be watchful? The definition of industry behemoths as described in national statistics encompasses many different product and geographic markets and the industry groupings are far too broad to properly assess market power. The observed trends could be reflecting the expansion of successful, efficient firms into ancillary product categories and new geographic regions for the benefit of consumers (Muris and Nuechterlein 2019). There is an important distinction between behemoths harming competitors and harming competition (Shapiro 2018). Effective competition by a firm is always tough on its competitors. Competition for a market is the key measure, not competition in a market. "It took several decades of diligent analysis from academics, practitioners, antitrust authorities, and courts to draw the critical distinction between harm to competitors and harm to competition" (Shapiro 2018). Behemoths, by competing for markets, drive innovation. "Competition for a market, rather than competition in a market, has produced some of the greatest innovation of the digital age" (Muris and Nuechterlein 2019). On the other hand, "When do we decide that low consumer prices are too much of a good thing and must be kept high to protect redundant jobs and avoid the challenge of retraining workers for more productive roles in the modern economy?" (Muris and Nuechterlein 2019).

Quality What is streaming and consolidation in this industry doing to content quality? Branded franchises, which comprise big-budget event movies, each creating a web of story lines, seem to be the most in demand.

"People say they want new ideas and fresh concepts, but in reality they most often go to the multiplex for familiar characters and concepts that remind them of what they already know they like. Big name brands like Marvel, Harry Potter, Fast & Furious, and Despicable Me consistently gross more than $1 billion at the global box office" (Fritz 2018).

Spielberg has voiced a similar concern: "A lot of studios would rather make branded, tent-pole, guaranteed box office hits … than take chances on smaller films. And those smaller films that studios used to make routinely are now going to Amazon, Hulu, and Netflix" (Barnes 2018).

This shift in tastes has occurred due to two important events: (1) the opening up of China and the Chinese consumer market and (2) the streaming revolution.

The streaming revolution has re-centered the creative impetus in television. "In 2016, networks and streaming services produced 454 original scripted series, more than double the number created in 2010" (Fritz 2018). Paradoxically, 2016 also saw the release of the largest number of franchise films, 37. Fritz writes portentously:

> The rise of original, risk-taking television is directly tied to the decline of original, risk-taking filmmaking and the dawn of the franchise age of film—one is which studios no longer coddle creative talent, release movies of every type for everyone, or pride themselves for taking risks on quality and new ideas. Instead, movie studios now exist primarily for the purpose of building and supporting branded franchises that continue in sequels, toys and theme-park attractions. (Fritz 2018)

This suggests that movie producers were choosing less risky strategies and were becoming more risk averse than in the hallmark era of Hollywood filmmaking like the 1970s.

The Entertainment Industry: Video Games

Like Pokemon Go, developed by Niantic, and unlike other content platforms, video gaming is an interactive platform with $140 billion in sales and considered the entertainment industry's biggest disrupter (Financial

Times 2019). The current leader in this gaming market, Fortnite, is developed by Epic Games, a social media platform.

Google recently announced a platform, Stadia, for video games that steers away from dedicated consoles toward cloud computing. This platform allows streaming of games to various devices including television, desktop computers and, notably, smartphones. Sony had a first-mover advantage in this content space with its PlayStation (1994) but is now developing a cloud-based game streaming application. Microsoft is also developing a similar service, xCloud, while Apple recently announced its own platform, Arcade.

Pioneers in the gaming market consisted of the millennials who populated the Sony PlayStation market and are "now raising their own children on the principle that gaming sits as an equal alongside books, music and movies." Even more critical for the growth of gaming is the presence of strong network effects. These are games requiring friends for online interactive war and are critically dependent on the size of the market (Financial Times 2019). The advent of 5G broadband, which could be standard within the decade, will facilitate the growth of this market by increasing both demand for interactive games and the supply.

The Entertainment Industry: Music

According to the 2018 report by the International Federation of the Phonographic Industry (IFPI), the US is the largest music market in the world and is projected to contribute $143.3 billion in value added to the US economy in 2016 or 0.8 percent of US GDP. The compound annual growth rate of this value added is 10.2 percent over 2012–2016, compared to the growth rate of US GDP which is 3.6 percent. The industry comprises a 1.3 percent share of US employment (US Music Industry: Jobs and Benefits 2018). At a state level, Tennessee, California and New York contributed most to their states' GDP in 2015, possibly reflecting regional tastes and technology.

While total digital consumption grew by 20.5 percent over 2016–2017, digital album and digital track sales fell by 19.6 percent and 23.4 percent, respectively. In one year! Meanwhile, total on-demand audio streams rose by 58.7 percent. Streaming has generated more revenue for the music industry than digital downloads, confirming the growing relevance of streaming technology in the music industry (US Music Industry: Jobs and Benefits 2018). Promoters of performing arts, agents/managers for artists and independent artists made up the largest category in music industry receipts ($54.6 billion in 2015) (*industry classification code 700–799*), followed by integrated record production/distribution, music publishers, sound recording studios and Internet broadcasting ($39.5 billion) (*industry classification code 500–599*). However, earnings per employee fell over the period 2012–2015 by the largest amount in these two categories (falling 1.7 percent in the former category and 1.6 percent in the latter category). Overall, earnings per employee fell 0.9 percent in the entire music industry over the 2012–2015 period (US Music Industry: Jobs and Benefits 2018).

US musicians only take home one-tenth of national industry revenues. Many artists blame the streaming model, which grants control to the platforms, such as Spotify and Apple. "Average per-stream payouts from the company are between $0.006 and $0.0084" (Wang 2018a). Well-known artists can make millions, while lesser known artists barely get by. Streaming is the disruptive technology in the entertainment world and the music industry has accepted it as the revenue generator.

Over the past year, Spotify is disintermediating the major labels, by striking licensing deals directly with independent artists. These contracts grant more cash flow and control rights to the artists. They are more like rental contracts where the license is not exclusive to Spotify, leaving the artist to re-contract with other platforms. Additionally, the revenue sharing is more equitable compared with that of the major labels—Universal, Sony and Warner. "By agreeing to a direct licensing deal with Spotify, artists and their representatives are able to keep the whole payout" (New York Times 2018).

Copyright, property rights over songs and albums, in the music industry is a complex tangle of arcane rules. The Music Modernization Act, signed into law by the president on October 11, 2018, will issue a blanket license, maintained by the US Copyright Office. This office will create a mechanical licensing collective which will maintain a database of all eligible works and will extend copyright protection to music created prior to the cutoff date under the old Copyright Act of 1976 (Copyright.gov 2019).

The blanket license will require streaming services to pay a regular royalty payment and audio recording engineers will be paid when the recordings are aired on both satellite and Internet radio (Wang 2018b).

However, plagiarism remains a serious unresolved issue in the "black box of royalties in the streaming era, a pit of unpaid money that hasn't yet made its way to artists because of faulty metadata or bad communication amongst the various services involved in reporting the proper numbers; it's worth has been estimated in the billions" (Wang 2018a). The creative side of a song is split between the audio recording of a song and the composition (lyrics plus melody). The recording copyright is usually held by the label, the firm that is bankrolling the artist.[6] When a song is purchased from iTunes, for example, that money goes to both composition and recording copyright holders. Similarly, when a song is streamed, or played on other commercial sites, both copyright holders get their predetermined fraction. Synchronization is the transaction whereby music is used in other digital content, such as television, film and radio (and YouTube), and these content producers pay a license fee, plus ongoing royalties. Satellite radio (AM/FM) pays only composition royalties, while Internet radio pays both kinds of royalties (Wang 2018a).

So, when you listen to a song on your car radio, only the writers of the song get paid not the artists behind the singer. When Bob Dylan sings "Big Yellow Taxi" on the radio, only songwriter Joni Mitchell, who wrote the song in 1970, gets paid.

Rap is the most widely heard poetry across the globe, in large part due to ICT. According to data from the consumer music profile from Recording Industry Association of America (RIAA), rap/hip-hop is the favorite music genre among music streamers (some of whom may have downloaded at least one track for free from a file-sharing service) and the second favorite music genre among paid subscribers to streaming services. Classic rock remains the most popular genre overall among all Internet music consumers (RIAA 2017b).

---

[6] Sound recording rights comprise performance rights on streaming services and radio; reproduction rights and sync rights for use on other media, such as television and film, and where possible YouTube.

Kendrick Lamar, a best-selling rapper, recently won the 2018 Pulitzer prize for music. "He's become a messiah for social progressives, a hero for wordplay-obsessed classicists, and a beacon for musical experimenters—in other words, he serves all the outlier groups" (Bradley 2017). Rappers, using rhythm, rhyme, clever wordplay and a system of symbolic meaning, rephrase ordinary language, giving it a heightened sensibility. Words enter the cultural fortress by influencing ideas and expanding our understanding of human experience with stories we might not otherwise hear. The beat is poetic meter transformed into audible form. In 2017, R&B/Hip-Hop became the most dominant musical genre, with seven of the top ten most consumed albums coming from that genre (Bradley 2017). The idea of giving voice to the experience of lifting yourself up from dire circumstances and experiences of injustice resonates worldwide.

However, more recently, producers, not artists, do much of the work, splicing together hook-heavy songs from a collection of digital files. Rappers write the words, then rhyme and record, letting the producer assemble them phrase by phrase. You have to cram far more syllables into a given bar in rap than you do in song. Consequently, there is a loss of variety—music tracks are remixing songs and creating joint productions between multiple artists and packaging old musical elements with minor new creations. As Jon Caramanica writes in *The New York Times*, "[J Cole and Kendrick Lamar's] victories underscore the clear self-segregating echo chamber that is hip-hop's mainstream, which relies upon and perpetuates narratives of triumph and conceitedness. For the most part, Mr. Cole and Mr. Lamar have worked against those values, and they've been punished for it" (Caramanica 2015). Neither of the two artists have had a Billboard Top 10 Single even while selling over a million copies of their albums and earning the platinum label from the RIAA.[7]

---

[7]To earn a platinum award from the Recording Industry Association of America (RIAA), the artist must have had over one million units of downloads or on-demand streams in the US, where 150 on-demand streams are considered equivalent to one download sale. To earn the gold award, the artist must have had over 500,000 units in the US only (Pollstar 2018). Billboard charts combine data from radio airplay and touring as well as social interactions on Facebook, Vevo, YouTube, Spotify and other popular platforms, in addition to sales and downloads.

This homogeneity in content is dangerous. New ideas can germinate when the ground is fertile, when the audience is receptive. The same old, same old is very soothing for risk-averse consumers, whose placidity revokes deviation from the norm. So new rap sounds much like old rap, with minor, insubstantial changes. There is no innovation in the arts. There can be little dispersion of opinion, need for discussion or debate over contrasting ideas. This can only lead to mental atrophy and diminished risk-taking.

John Pareles writes in *The New York Times* that Tierra Whack's album Whack World with its 15 one-minute songs was a "miniature tour de force" and "a brilliant strategy for combat in the attention economy: keep things short and visual (Pareles 2018). Because, really, who has time for music?" Music competes with other screen modes of entertainment since streaming is keyed to the screen. So email, social media, text messages, notifications and so on all compete on this tiny screen. Artists themselves are inundated with multiple demands, beyond the studio—marketing, promotion and touring. The need to constantly stay on the listeners' radar, as the

> internet looms as a giant, insatiable maw, constantly demanding more content with less payback. YouTube and streaming have made each song a click rather than a purchase. ... In that endless stream, the idea that a song is a thought-out, carefully distilled utterance was bound to erode.
>
> Add another factor: social media. The initial promise was increased connectivity, a way to reach and respond to fans quickly and candidly: no gatekeepers, no filters. But for most pop aspirants, the result has been the end of creative seclusion. There's pressure to keep offering new material, musical and nonmusical, making the promotional cycle as endless as the internet workday. Maybe it's a song snippet or a glimpse of a video shoot. Or maybe it's a Twitter feud, a media-ready screed on Facebook or a fashion experiment on Instagram. (Pareles 2018)

Live Events

Streaming music provides the impetus for live performances as listeners want to engage at a deeper level with their favorite performer. While streaming hits sound, "unfinished" live music performances are much in demand as evidenced by ticket prices. Gross revenues globally, according to concert-promotion magazine *Pollstar*, were $5.6 billion in 2018 (Pareles 2018).

Streaming is whetting the public's appetite for live performances by the artists, so tours and music festivals are dominating the scene. Ticket prices are at an all-time high. The concert company, Pollstar, finds that as of June 2018 average ticket prices were $96.31. The highest price of $509 was for "Springsteen on Broadway." Interestingly, most of the lower bowl, higher-priced seats in stadiums were filled to capacity, while the less expensive upper seats were often unsold, even in the secondary market (Pollstar 2018). Of the top ten grossing tours in 2018, Ed Sheeran dominated with $2.624 million in ticket sales, and Disney-on-Ice ($1.095 million), Bruno Mars ($0.739 million), Pink ($0.691 million), Kenny Chesney ($0.669 million) and The Rolling Stones ($0.633 million) followed, respectively.

In the sports market, the smartphone era has ushered in new ways of watching live games. Traditional broadcast television is replaced by bite-sized highlights on YouTube or clips made by friends watching the games in a Russian stadium. Consequently, "watching live matches will be just a fraction of millennials' engagement with this World Cup" (Kuper n.d.).

At some point these various streams of communication, information and entertainment meet in a gathering storm of content tsunami.

References

Aguiar, Luis and Joel Waldfogel, Netflix: global hegemon or facilitator of frictionless digital trade? *Journal of Cultural Economics*, November 2017.

Barnes, Brook. 12/16/2018. *Netflix's Movie Blitz takes Aim at Hollywood's Heart.* Accessed 4/2/19 from https://www.nytimes.com/2018/12/16/business/media/netflix-movies-hollywood.html

Baumeister, Roy E, Ellen Bratslavsky, Mark Muraven, and Dianne M. Tice. n.d. *Ego Depletion: Is the Active Self a Limited Resource?* Accessed 3/12/18 from https://faculty.washington.edu/jdb/345/345%20Articles/Baumeister%20et%20al.%20(1998).pdf

Bradley, Adam. 2017. *Book of Rhymes: The Poetics of Hip Hop*, New York: Basic Books.

Brandom, Russell, *ATT's Time Warner merger kicks off a new era of streaming-video monopolies*. Accessed 6/14/2017 from https://www.theverge.com/2018/6/12/17456014/att-time-warner-merger-antitrust-streaming-video

Bureau of Labor Statistics: Labor force projections and highlights: 2016–2026. *Monthly Labor Review, 2017*. Accessed from https://www.bls.gov/opub/mlr/2017/article/projections-overview-and-highlights-2016-26.htm

Bureau of Labor Statistics: Beyond the Numbers. April 2018. Retrieved from https://www.bls.gov/opub/btn/volume-7/pdf/so-you-want-to-be-in-pictures.pdf

Caramanica, John. 2015. *Vince Staples and J. Cole, Outsiders in the Middle of Hip-Hop*. Accessed from https://www.nytimes.com/2015/07/12/arts/music/vince-staples-and-j-cole-outsiders-in-the-middle-of-hip-hop.html

Copyright.gov. 2019. Accessed 4/9/2019. from https://www.copyright.gov/music-modernization/

Financial Times March 23, 2019. Accessed 3/23/19 from https://www.ft.com/content/7be7cc00-4c9d-11e9-bbc9-6917dce3dc62

Fritz, Ben. 2018. *The Big Picture: The Fight for the Future of Movies*. New York: Houghton Mifflin Harcourt. March 2018.

Gomes-Casseres, Benjamin. 2018. What the Big Mergers of 2017 Tell Us About 2018. *Harvard Business Review*, January 2. Accessed 8/30/2018 from https://hbr.org/2017/12/what-the-big-mergers-of-2017-tell-us-about-2018

IFPI. Accessed 4/3/2019 from https://ifpi.org/facts-and-stats.php

Koblin, John. 2017. *Netflix Says it will spend $8billion in content next year*. Retrieved from https://www.nytimes.com/2017/10/16/business/media/netflix-earnings.html?_r=0

Krause, Elinor and Isabel Sawhill. 2016. *How free time became screen time*. Brookings Memo. Retrieved from https://www.brookings.edu/blog/social-mobility-memos/2016/09/13/how-free-time-became-screen-time/

Kuper, Simon, *World Cup: why millennials will transform football's greatest event*. Accessed 6/15/2018 from https://www.ft.com/content/fd9c10b8-6a70-11e8-b6eb-4acfcfb08c11

Muris, Timothy and Jonathan Nuechterlein. 2019. *Anti-Trust in the Internet Era: The Legacy of US v A&P*. Electronic copy available at: https://ssrn.com/abstract=3186569

New York Times, 9/6/2018. Accessed 9/6/18 from https://www.nytimes. com/2018/09/06/business/media/spotify-music-industry-record-labels. html?dlbk=&emc=edit_dk_20180906&nl=dealbook&nlid=20718461_ dk_20180906&te=1

Pareles, Jon. Dec 26, 2018. Pop in the Era of Distraction. *New York Times*. Accessed 12/28/2018 from https://www.nytimes.com/2018/12/26/arts/ music/pop-music-social-media-distraction.html?em_pos=large&emc=edit_ ms_20181228&nl=louder&nlid=20718461edit_ms_20181228&ref=h eadline&te=1

Pollstar, 2018. *Mid-year special features*. Accessed from https://www.pollstar. com/article/2018-mid-year-special-features-top-tours-ticket-sales-business- analysis-135890

RIAA. 2017a. Accessed on 5/21/19 from http://www.riaa.com/wp-content/ uploads/2018/03/RIAA-Year-End-2017-News-and-Notes.pdf

RIAA. 2017b. Accessed 9/6/2018 from http://www.riaa.com/wp-content/ uploads/2018/05/MusicWatch-Consumer-Profile-2017.pdf

Shapiro, Carl. 2018. *Antitrust in a Time of Populism*. International Journal of Industrial Organization. 61. Accessed on 5/20/2019 from https://www.sci- encedirect.com/science/article/pii/S0167718718300031?via%3Dihub

Statista. Accessed 4/3/2019 from https://www.statista.com/chart/15763/most- popular-music-genres-worldwide/

Tuftci, Zeynep, "*It's the Democracy-Poisoning Golden Age of Free Speech*," Accessed on 5/20/2019 from https://www.wired.com/story/free-speech-issue-tech- turmoil-new-censorship/

US Music Industry: Jobs and Benefits. 2018. Accessed 9/4/2018 from http:// www.riaa.com/wp-content/uploads/2018/04/US-Music-Industries-Jobs- Benefits-Siwek-Economists-Inc-April-2018-1-2.pdf

Wang, Amy. 2018a. How Musicians Make Money – Or Don't at All – in 2018. *Rolling Stone Magazine*, August. Accesses 4/9/2019 from https://www.roll- ingstone.com/music/music-features/how-musicians-make-money-or-dont- at-all-in-2018-706745/

Wang, Amy. July 2018b. Why Your Favorite Concerts are Bigger and More Expensive – Than Ever. *Rolling Stone Magazine*. Accessed 4/9/2019 from https://www.rollingstone.com/music/music-features/why-your-favorite-con- certs-are-bigger-and-more-expensive-than-ever-699722/

6

Content Tsunami and the Attention Deficit

Cognition is about thinking, about connecting ideas and then retaining them. A tsunami of diverse and noisy content cannot connect or form "multiple associations with every fact we care to retain" due to its very diversity (James 2019). Proliferation in content across multiple categories—communication, information and entertainment—and dispersion in quality within each category creates noisy jitter instead of coherent content.[1] Video and audio content have compounded programming choices due to streaming technology enabling multiple media platforms. Noisy content soaks up attention, leaving less mind-space for creating "enterprises of great pith and moment" as Hamlet articulates in William Shakespeare's play, (see epigraph in the preface). Today, Trevor Noah says, "Ain't nobody got time for that."[2]

It has been acknowledged for some time that the mere presence of a smartphone diminishes focus, making work disordered and thereby lowering productivity, an issue raised in Chap. 2. Researchers have found that "cellular phone notifications alone significantly disrupted performance on

---

[1] Jitter is a technical term in network communications and refers to the variability in data flow between systems. Jitter can result from network congestion and route alteration.

[2] The commentator Trevor Noah labels important issues that have been swept under the public radar as "Ain't Nobody Got Time for That" in his popular *Daily Show*, airing on *Comedy Central*.

© The Author(s) 2019
S. Bhatt, *The Attention Deficit*, https://doi.org/10.1007/978-3-030-21848-5_6

an attention-demanding task, even when participants did not directly interact with a mobile device during the task. The magnitude of observed distraction effects was comparable in magnitude to those seen when users actively used a mobile phone, either for voice calls or text messaging" (Storhart et al. 2015). What is less well recognized is the impact of content transmitted by this device.

In this chapter, I generalize digital content to include news. The focus in Chap. 4 was on media and entertainment content, while this chapter adds news and communication in the form of information about the world. Henceforth, digital content is about the entire universe of communication, information, media and entertainment or CIME.

Noisy Content: Error Code 13 and 14—Invalid Data and Insufficient Storage

Problems with information are examined in the economics literature in two distinct ways: at the source level and at the recipient level. First, at the source level, information is defined as *incomplete* when it is fragmented and not all relevant information is available. Further, information can be *imperfect* when some group has more information relative to another, in a situation called asymmetric information. These informational asymmetries arise when, for example, the employee is better informed about her work effort compared with the information collected by her employer. Individuals solve this incompleteness and imperfection by incurring "frictions" or search costs, which are the costs of acquiring and processing information. Closely related is "rational inattention," the idea that people acquire information based on a cost-benefit calculation. For example, if I were looking to buy a car, I would balance the costs of obtaining information about prices and car features against the pleasure from driving a particular car at the best price. But, this sort of computation requires awareness of the information required, and this can be difficult when the information set itself is unknown or ambiguous. It follows that the costs of acquisition may also be unknown.

Second, at the recipient level, individuals can have *"mental gaps or psychological distortions in information-gathering, attention and processing [italics mine]"* (Handel and Schwartzstein 2018). In financial markets, this would reflect limited financial literacy. In insurance markets, people exhibit limited understanding of basic probabilities. People overestimate the likelihood of rare events and overweight rare events in their decision-making (Kahneman 2011). The misinterpretation of facts when individuals succumb to the availability heuristic—a recent spate of car crashes triggers an immediate fear of road travel—is an example.

My idea of noisy content is not concerned with either of these two fault lines in information access and data processing. I am addressing the issue of excess and involuntary data flow, such that the brain signals a system error. The digital analogy from the Windows operating system is "Error Code 13: The data is invalid" and "Error Code 14: Not enough storage is available to complete this operation" (Microsoft 2019).

Simplified information processing or cognitive ease, even in the absence of any cognitive overload, is an inherent human characteristic. "The sophisticated allocation of attention has been honed by a long and evolutionary history" according to Kahneman. Importantly:

> A general "law of least effort" applies to cognitive as well as physical exertion. The law asserts that if there are several ways of achieving the same goal, people will eventually gravitate to the least demanding course of action. … Laziness is built deep into our nature. (Kahneman 2011)

Physicians, overloaded with data from electronic health records (EHR), often ignore critical information amidst this data tsunami. According to Verghese:

> [I]n an I.C.U., a blizzard of monitors from disparate manufacturers display EKG, heart ate, respiratory rate, oxygen saturation, blood pressure, temperature and more, and yet none of this is pulled together, summarized and synthesized anywhere for the clinical staff to use.

Furthermore:

> [A] clinician will make roughly 4000 keyboard clicks during a busy 10-hour emergency-room shift. In the process, our daily progress notes have become bloated cut-and-paste monsters that are inaccurate and hard to wade through.

As a consequence, the \$3.4 trillion health-care system can't prevent over 250,000 deaths per year due to medical error, mostly resulting *"from failing to listen to the story and diminishing skill in reading the body as a text [italics mine]"* (Verghese 2016).

Error Code 13: Invalid Data Infringes on Private Mental Space

Media behemoths distribute content, an increasing fraction of which is user-generated information, manipulating users' attention to create addiction to the content. By priming or suggesting various other videos or automatically playing related videos, YouTube promotes associative thinking. These strategies influence our thoughts as manipulating context can alter thinking and behavior. For example, familiarity arising from repetition or frequency is a form of priming. If I suggest a phrase or image that is familiar and invokes good feelings, then I have primed you to think and behave positively. Movie sequels and collaborative song tracks, that draw upon the common well of music fragments, generate this kind of fluency or cognitive ease and hence capture the audience. Mainstream hip-hop, for example, as understood mainly through the radio, has a fairly uniform dominant narrative which rests upon triumph and pride.

Fake news and fake images distort our reality, and even worse, when we aren't sure of what to believe, lead us to distrust our very senses. "Fabricated videos will create new and understandable suspicions about everything we watch." Take virtual reality technology which purports to transport us to another world, by playing upon our sense of perception. Can consumers be manipulated by VR? "Studies have already shown how VR can be used to influence the behavior of users after they return to the physical world, making them either more or less inclined to altruistic behaviors" (Foer 2018). More insidious is content that is computer generated, as in fake images. AI can blend familiar faces onto pornographic scenes in a new genre that is "one of the cruelest, most invasive forms of identity theft invented in the internet era. At the core of the cruelty is the acuity of the technology: A casual observer can't easily detect the hoax" (Foer 2018).

The premise of immersive technology is the ability for creators to completely curate a person's visual and auditory experience. It has the potential to transform any activity that relies on these senses (Verghese 2016). Virtual reality and augmented reality, therefore, can impact our thoughts and ideas and, therefore, our values which constitute our criteria for judgment. This translates into behavioral changes. AR will change how people work and learn but also experience art and stories (entertainment).

Two computer scientists in Spain, Mel Slater and Maria Sanchez-Vives find that "wearing a different body does have profound psychological effects. … When white people are embedded in a black body, says Slater, they start to show marked decreases in measures of implicit racial bias—an effect that lasts at least a week after exposure." A VR treatment called SnowWorld for pain control in people with severe burn injuries reduces mental images of pain by immersing patients in glacial canyons and playing with snow (Waldrop 2017). On a more pragmatic level, research based on randomized control studies has shown the success of VR exposure therapy for anxiety and pain in a Level I Trauma Center (Martin et al. 2018).

How much attention or cognitive bandwidth is available to think freely when we are confronted with a tsunami of invalid data? There are two issues here. First, when distorted data such as misinformation and radical ideas are directed toward particular individuals, there is invasion of personal property in the sense of compromised thinking or personal mind-space. Second, this data might be molding our ideas, which constitute the foundation of our cultural scaffold (see Appendix to Chap. 1 for a discussion of culture). This steps at the very front door of free will.

Privacy

With connectivity comes information transfer, some of which may not be voluntary. The public focus has mainly been on the involuntary revelation of information, but it is equally important to recognize the involuntary receipt of information, which I call intrusion on private space. Personal property extends beyond tangible records to include one's attention or mental resources. Theoretically, property rights should also extend to personal mind-space.

Personal property includes tangible possessions such as one's home and personal effects, personal and financial records and cell-site location information. The US Supreme Court ruled in June 2018 that the government's use of the defendant's cell phone records was a violation of Fourth Amendment rights. This ruling augmented the interpretation of privacy to include more than property-based privacy expectations. By voluntarily disclosing location to the wireless carrier, a person is sharing their data. However, since a phone is always on a person, a detailed record of the person's movements is also revealed. This is not intentional and therefore not voluntarily disclosed.

Not only is property protected under this right, but any reasonable expectation of privacy is also granted protection against any intrusion and search, requiring a warrant (in 2012 *U.S. v. Jones* 565 U.S. 400 did not admit global positioning system [GPS] tracking data as admissible evidence). At the same time, information voluntarily shared with third parties is not granted this 4th Amendment protection (in 1976, *U.S. v. Miller* 425 U.S. 442 held that financial records turned over to a bank did not guarantee privacy). In 2018, *Carpenter v U.S.* 585 the US Supreme Court held that "historical cell-site records present even greater privacy concerns than the GPS monitoring in Jones: They give the Government near perfect surveillance and allow it to travel back in time to retrace a person's whereabouts." CSLI, cell-site location information, is neither voluntarily shared nor limited in scope like business records, but is "an exhaustive chronicle of location information casually collected by wireless carriers" (SCOTUS 2017).

There are diverse forms of national organization and the use of data by some large, authoritarian states, such as China, to maintain stability is quite possibly a matter to be determined by those directly affected. Safety in a large country like China is perhaps only ensured by monitoring the population. On the other hand, those being surveilled by complex facial recognition technology to assign credit scores may not be aware of the trade-offs made when they acquiesce in this large-scale parenting-like process.

At a granular level, there is the infiltration of individual mind-space by fringe groups seeking to destabilize nations by promoting their own brand of incendiary beliefs. Citing Jihadism, the perpetrators of the mass shooting in San Bernardino, California, destroyed 14 lives and harmed

22 others, on December 2, 2015, having just posted allegiance to The Islamic State of Iraq and Syria (ISIS) on Facebook. Earlier that year, on November 13, 2015, terrorist attacks killed 130 people and injured 413 in Paris, France. ISIS claimed responsibility.

Federal Bureau of Investigation (FBI) investigators concluded that extremist ideology acquired via the Internet as well as travel to Saudi Arabia prior to the attack was responsible. James Comey, then director of the FBI said in testimony before a Senate Judiciary Committee on December 9, 2015, that the Islamic State sought to attract individuals and "radicalize in their home" so that these "home grown radical extremist to [sic] kill innocent people on behalf of a foreign terrorist organization" (Comey in Congressional Testimony 2019).

Communication need not be overt as in email or text messages. Subtle persuasion through an unsuspected, benign mechanism can be consequential as well. As Comey said to the Senate, "We have seen a number of cases in which subjects of investigations have communicated through gaming channels, either through live action games or sometimes through app games on devices."

Many thought leaders make the case that social media have violated our basic property rights over privacy and that face-to-face interaction is superior to digital contact. *The New York Times* has launched The Privacy Project, which will no doubt initiate public dialogue, with some arguing that although there is no clear right to privacy in the US Constitution, the UN has recognized privacy as a basic human right, a response to government surveillance (United Nations Human Rights 2019). We are, in fact, creating a private surveillance state when we install home security cameras that ostensibly "watch" our children or protect the front door, but what about the privacy of domestic workers or the postman? Anxious neighbors may post these faces on the neighborhood social network inducing racial profiling in sanitized, closed neighborhoods.

Accidental sharing of data can have adverse consequences, as when financial data are divulged involuntarily. Unintended disclosure includes sensitive information posted publicly, mishandled or diverted to a mistaken party via email. In a breakdown of data breaches by breach type, 24.3 percent of all data breaches in 2017 involved unintended disclosure of financial information, but a much larger fraction of records was lost by this mistake—84.7 percent of the total records were lost via mistaken disclosure (Privacy Rights Clearing House 2019). In the last two years, two of the top five cases under investigation for breaches were due to unauthorized access or disclosure. A case in point is the mishandling of health-care records of 1.25 million individuals in the Employee Retirement System of Texas in October 2018 (US Department of Health and Human Services, Office for Civil Rights 2019). Finally, a recent study found that accidental sharing of records occurs primarily at educational institutions and government and military organizations (Rogers 2019).

There has already been much discourse on privacy.[3] None have addressed the question of the value of privacy. Is it a private consumption good which is purchased at the appropriate market-determined price? Or is it a public good where the common value exceeds individual private value and therefore collective action is vital? In the first case, individuals would be hard pressed to assess their value of privacy since it is context dependent, changing with the circumstances.

Tim Wu says "mass privacy is the freedom to act without being watched and thus to be who we really are—not who we want others to think we are" (NYT 2019). Another more realistic view is that with social media we are in the world of the looking glass self, a concept created by the American sociologist and founding member of the American Sociological Association, Charles Horton Cooley in 1902. The looking glass self is the idea that our self-image is shaped by our connections with others. Sharing doesn't mean narcissism but rather careful shaping of one's presentation depending upon the context. Imagining how we appear to the world and the world's judgment of that appearance is predicated upon the criteria selected for that judgment.

Shaping of Ideas

Well before digital communication became a standard mode of connectivity, advertising was creating many of the same concerns that today we

---

[3] See, for example, Cal Newport, Digital Minimalism: Choosing a Focused Life in a Noisy World and Shoshana Zuboff's Surveillance Capitalism (Newport 2019; Zuboff 2019).

associate with social media. The advertising model that is the focus of much criticism of digital media has attracted the same kind of interest since Richard Pollay wrote that "[w]e are all potential victims of the invidious comparisons of reality to the world seen in [social media] advertising. Once convinced that the grass is greener elsewhere, one's own life pales in comparison and seems a life half-lived" (Pollay 1986).

It was also acknowledged that there were consequences to confronting the imagery in the looking glass held up by advertising. Priming has been demonstrated to "threaten our self-image as conscious and autonomous authors of our judgments and our choices" concluded Kahneman in 2002 (Kahneman 2011). The notion that our very ideas were being molded was also recognized by Pollay, who wrote that "commercial persuasion appears to program not only our shopping and product behavior but also the larger domain of our social roles, language, goals, values, and the sources, of meaning in our culture" (Pollay 1986).[4] Consumption was romanticized by transforming our psychological desires and needs into consumption, making objects more important than people. Commoditization of relationships, as discussed in Chap. 3, is the modern social media version of exactly the same phenomenon. Social media is like the cultural mirror of advertising that upholds behavioral norms patterning the good life. And, by comparison, our own life is mediocre.

In the work culture, individuals who identify with an organization perform better than those who perceive themselves as outsiders. Akerlof has written about influence and the molding of identities to elicit a better fit, and hence superior performance, in organizations. Ideally, in the work environment, it is optimal to allocate workers to a job with which they identify and is consistent with their self-image prior to any job-related interaction. The US Army training program at West Point aims to influence these choices. "The Army's aim at West Point is to change cadets' preferences. They wish to inculcate non-economic motives in the cadets so they have the same goals as the U.S. army. Alternatively stated, the goal of West Point is to change the identity of the cadets" (Akerlof and Kranton 2005).

---

[4] Macroeconomists would agree with Pollay that "[A]dvertising [social media] is seen as inducing us to keep working in order to be able to keep spending, keeping us on a treadmill, chasing new and improved carrots with no less vigor, even though our basic need may well be met" (Pollay 1986).

In general, changing workers' identity from being outsiders to being insiders allows the organization to substitute nonmonetary incentives for wages and salary. When there is uncertainty on the job, workers' effort is hard to gauge and high effort at peak times is critical to the firms' success. In such situations, it is less costly to elicit effort from an in-group employee than to motivate an outsider (Akerlof and Kranton 2005). Management styles that create distance by, for example, strict enforcement of rules, inculcate a sense of us-versus-them among workers who are correspondingly incentivized to shirk. In volatile times, when demand shrinks, these workers remain unmotivated to put in imagination and extra effort to save the day.

Distinct from content that is obviously advertisement is digital content in the broader communication, media and entertainment industries. Can this larger body of content achieve the same framing as advertising is doing at the commercial level?

Mutual consensus, which is sharing beliefs and ideologies, is the cement that strengthens connections and thus builds self-esteem. Consensus also provides a sense of security, which perhaps as a group uncertainty can be abated. Also called shared reality, consensus formation with a group or person is achieved by absorbing the beliefs and norms of that other group or person (Sinclair et al. 2014).[5] There is strong pressure to "like" the same movies, songs or YouTube videos. Even as there is no true understanding or appreciation for the content, momentum for consensus building drives acquisition of shared tastes. Popularity of content, then, is not just because the content is objectively top ranked but rather consensus driven, explaining the phenomenon of network effects discussed in Chap. 5.

People become more like one another by this process of preference adoption of the group. This creates homogeneity of preferences in the demand for content, especially entertainment, and suppliers accommodate by making high-action, low-dialogue movies that can be neatly summarized, since production costs are lower. The high-concept model adopted by the movie production industry (see Chap. 2 for details on high-concept films) is an illustration of this trend. US distributors hold

---

[5] Stacey Sinclair calls the adoption of group norms affiliative social-tuning, when driven by self-esteem building, and epistemic social-tuning when driven by a desire for certainty (Sinclair et al. 2014).

the top rank in box office revenues for the past 30 years, among all the English language movie distributors worldwide. The highest grossing movie worldwide, in any language, is the US production, Avatar (Box Office Mojo 2019).

Error Code 14: Insufficient Storage Compromises Risk-Taking

Is the application of sheer mental effort sufficient to counter these changes in perception and behavior? Mental effort or cognitive bandwidth is the number of hours that the mind can reasonably devote to process data.[6] Cognitive stress arises when a lot of data are presented requiring high cognitive bandwidth. In contrast, cognitive ease or fluency is a state with low data rates, requiring fewer hours of mental effort.

Baumeister (Baumeister et al. n.d.) explains, "*The pattern of ego deple-tion suggests that some internal resource is used by the self to make decisions, respond actively, and exert self-control [italics mine]*" and this resource is cognitive bandwidth. Moreover, once a train of thought has been invoked by some digital content, it unconsciously releases associated thoughts, making it even harder to curb the distraction. To actively monitor and curate incoming information or content, we utilize cognitive bandwidth. This process impairs the very judgment necessary to effectively separate the noise from true content.

Tim Berners-Lee led the second digital revolution by the creation of World Wide Web in 1993, which was propelled by the launching of Google in 1998 and Wikipedia in 2001, but this force did not last beyond the decade. Technological advancements have led to the displacement of labor by machines, but with limited reinstatement of new tasks, new job categories (Acemoglu and Restrepo 2018). Why has it been so difficult to

---

[6] The brain of a *computer* is the central processing unit (CPU) and CPU speed is roughly the num-ber of calculations performed in a second; it is measured in frequencies, so a CPU speed of 2.2 GHz performs 2,200,000,000 operations per second. Bandwidth is a measure of how fast the *network* can transmit data. Bandwidth is measured in bits per second or in terms of the width of the fre-quency band, the range of frequencies available (3000 Hz). It is the latter analogy that I am drawing upon when defining cognitive bandwidth—the range of hours available. Latency is the actual time it takes for data to go through the network, so constriction of cognitive bandwidth is latency (SCOTUS 2017).

expand the technological possibilities offered by the universal digital representation of data to previously unimagined uses? What is holding entrepreneurs back?

How does one explain Robert Gordon's results on the growth deceleration in US labor productivity from $2.26 of output per hour during 1994–2004 to $1.38 in the following decade, 2004–2014, and with a projected further decline to $1.2 from 2015 to 2040? Gordon says that, first, economic growth is not a steady process and second, that "economic growth since 1970 has been simultaneously dazzling and disappointing. This paradox is resolved when we recognize that advances since 1970 have tended to be channeled into a narrow sphere of human activity involving entertainment, communication, and the collection and processing of information" (Gordon 2016).

Fiduciary responsibility to shareholders and short-term horizons mandated by financial analysts obligate managements to prioritize shareholder (quite possibly all stakeholder) value. So one possible answer is behemoth firms are pinching off the grass shoots of startups in the dynamic game of strategic competition.

The low interest rate environment prevailing since 2000 has played an important role in reducing the incentive to invest (see Figs. 6.1 and 6.2). Liu, Mian and Sufi propose an explanation, which they confirm using stock price data from 1980 to 2018, for the slowdown in investment and economic growth even as long-term interest rates have declined (Liu et al. 2019). Behemoths respond strategically to low interest rates by investing in productivity-enhancing technology in order to maintain the productivity gap between themselves and more sluggish firms. Low-productivity firms facing an insurmountable barrier to catching up with the leaders give up and average productivity in the industry falls, widening the gap such that market concentration increases. Behemoths sprint ahead faster by deepening their technological capabilities when interest rates and borrowing costs slide downward.[7] Laggards exhibit the "discouragement effect," when the marginal value of catching up is not worth the cost of investment. Laggards then end up being the granular, small firms, whose

---

[7] Bhatt (2017) calls this "Red Queen" competition, after the character in Lewis Carroll's *Through the Looking Glass*, who says "Now, here, you see, it takes all the running you can do, to keep in the same place." Even the leaders have to sprint faster to simply maintain their leadership position (Bhatt 2017).

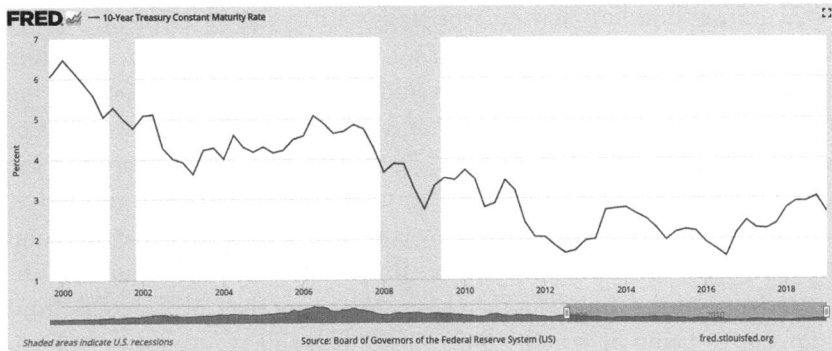

Fig. 6.1 Ten-year treasury constant maturity rate, quarterly average rates from 1999 to 2019. (Source: Board of Governors of the Federal Reserve System (US), 10-Year Treasury Constant Maturity Rate [DGS10], retrieved from FRED, Federal Reserve Bank of St. Louis. Accessed 4/5/2019 from https://fred.stlouisfed.org/series/DGS10/)

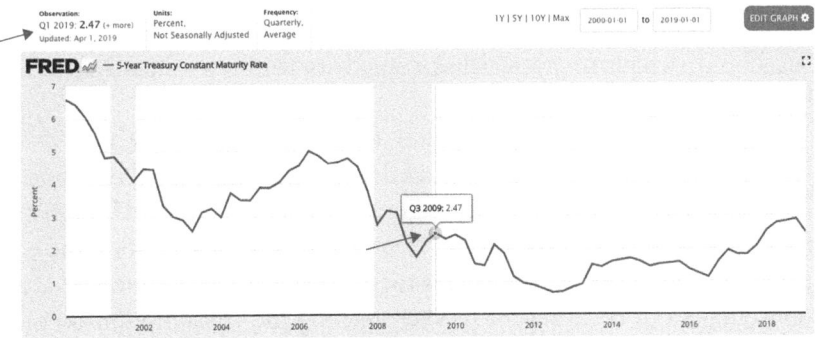

Fig. 6.2 Five-year treasury constant maturity rate, quarterly average rates from 1999 to 2019. (Source: Board of Governors of the Federal Reserve System (US), 10-Year Treasury Constant Maturity Rate [DGS10], retrieved from FRED, Federal Reserve Bank of St. Louis. Accessed 4/5/2019 from https://fred.stlouisfed.org/series/DGS10/)

goal is survival rather than growth. But the leaders end up becoming "lazy monopolists" when the gap has widened sufficiently to cement their advantage. In fact, overall productivity and economic growth stagnates. As discussed earlier and in Chap. 1, productivity has fallen in almost all advanced countries, starting well before the Great Recession, and has been persistent.

Acquiring budding new enterprises is part of this investment strategy by the leaders, which might explain some of the startup deficit. A decline in interest rates promotes mergers and acquisitions on the part of the behemoths, which absorb newly founded firms due to the lower cost of investing. Venture capitalists (VCs) "now talk of a kill-zone around the giants. Once a young firm enters, it can be extremely difficult to survive. Tech giants try to squash startups by copying them, or they pay to scoop them up early to eliminate a threat." But these "early 'shoot-out' acquisitions have sapped innovation," changing the motives of founders from growing to being acquired. "Since 2013 [Alphabet] has spent $12.6 b investing in 308 startups" (American tech giants are making life tough for startups 2018).

There are counterexamples such as Snap's refusal to be acquired by Facebook so that it could develop a unique texting platform which has worked to its benefit. Twenty-seven percent of US adults use the platform according to a survey done in January 2018 by Pew Research Center (see Fig. 2.1 in Chap. 2). But Snap is losing users to Instagram (owned by Facebook), which many feel has copied its features—Snapchat Stories became Instagram Stories.

Why isn't there a more compelling force driving inventiveness?

At a deeper level, the answer lies in the decoupling of time from content. This detachment lies at the heart of the attention economy and its signature feature—the cognitive bandwidth deficit. The freedom to consume content when and where we wish leads to excess consumption since the attention budget can be disregarded when making the consumption decision. Numerous avenues for information and communication can be superficially explored without realizing that there is a time and attention constraint precluding deeper engagement. This kind of over extension is similar to the mind-bending effects of scarcity.

Labeling this the scarcity mindset, Mullainathan and Shafir show, through experiments and field research, that "*scarcity captures the mind [italics mine]*" and alters how we think, usurping our "bandwidth" or mental capacity by focusing on exactly that which is scarce, ignoring everything else. Mullainathan and Shafir call this reduction in mental capacity the bandwidth tax on cognition, making individuals less insightful, less forward looking and controlled (Mullainathan and Shafir 2013).

In the current context, tunnel vision or selective filtering of information arises when the mind focuses exclusively on the fact that there is much content to consume and little time to execute. The two attributes of cognition—executive control and fluid intelligence—are compromised when the available capacity is significantly less than that which is strictly necessary or perceived as strictly necessary. The presence of "slack" or spare mental capacity reduces the negative consequences of the bandwidth tax. "Scarcity not only raises the costs of error; it also provides more opportunity to err, to make misguided choices" so that slack in mental capacity gives us "room to fail" and that in itself reduces the tax (Mullainathan and Shafir 2013).

There are two kinds of scarcity: objective scarcity and perceived scarcity. When the scarcity is about poverty, resources fall short of that necessary for mere survival, so this is an objective scarcity. When the scarce resource is time, the multiple demands on time include optional, luxury consumption (leading to the equivalent of perceived scarcity) as well as necessary consumption of content for daily living (creating objective scarcity). For instance, a labor shortage of truckers will likely raise prices of goods that need to be trucked from out-of-state and long distances so if I were a local baker, with neighborhood clients, my profits could be negatively impacted by the increase in input costs, like flour and sugar. Knowledge of this impending shortage would assist me in planning production, so I must make time for this information.

Compromised bandwidth has implications for dealing with uncertainty in everyday affairs. There is risk, and then there is ambiguity. Risk is uncertainty within a known framework, with well understood outcomes and their probabilities. One can purchase home insurance to mitigate huge financial losses from fire or burglary, and hurricanes. Listing all future outcomes and discounting them appropriately provides a present value of all possibilities. The standard way of discounting, exponential discounting, applies a discount rate that is constant and provides the current dollar value equivalent of future dollars. More realistically, recognizing that individuals abhor delayed gratification in the immediate future compared with delayed gratification in the distant future, hyperbolic discounting applies a larger discount rate to immediate outcomes. Waiting an extra day now for a reward of $100 is more painful than waiting an extra day one year from now for the same reward. When context matters, discount rates may change as the environment changes.

Ambiguity is uncertainty multiplied across unknown frameworks, unimagined outcomes with unfathomable probabilities.[8] How does one insure against what one doesn't know? The ability to comprehend the nuances of ambiguity is what distinguishes human intelligence from machine intelligence (see the conclusion for a detailed discussion).

With invalid data crowding our mental space, there is insufficient capacity, a cognitive shortage. Attempts to organize and classify data itself use scarce bandwidth. A picture accompanied by a simple model clarifies the thesis.

Model of Attention Deficit

This model provides a simple, but novel, framework for visualizing how the interplay of content tsunami, artificial intelligence and effortful filtering jointly creates an attention deficit. Attention is represented by rationally chosen cognitive bandwidth hours (CBH) denoted by h_a such that the hours of attention plus the hours of leisure, including sleep, add up to the daily constraint of 24 hours: $h_a + h_l = 24$ (as discussed in Chap. 4). This individually rational calculation does not account for the externalities emanating from social and entertainment media—sharing may have a positive effect on the individual but collectively it becomes a content tsunami.

Unpacking the impact on each person, there are two conflicting forces: desired cognitive bandwidth hours and available cognitive bandwidth. Desired cognitive bandwidth hours are the hours required, by any given individual, to consume a given quantity of content, where time can be devoted to valid, useful content, requiring θ_{good} hours, or the tsunami of fake and misleading content, utilizing $\theta_{tsunami}$ hours. Hence, desired cognitive bandwidth hours, which can be interpreted as the demand for attention, can be formulated as:

$$\text{Demand for attention} = \theta_{good} + \theta_{tsunami}$$

---

[8] Uncertainty can be categorized as (1) risk or uncertainty within a model—we are not sure about what parameters to use for each model and we have uncertain outcomes with known probabilities, (2) ambiguity—we are not sure what natural science or economic model to use; there is ambiguity across models with unknown weights for alternative possible models and (3) misspecification—in which we are concerned that our models are misspecified in various ways and we have uncertainty about the models and unknown flaws of approximating the models.

Available cognitive bandwidth is a measure of an individual's functional capacity or how fast the data can be processed. (In communications systems, bandwidth is a measure of how fast the network can transmit data, the data rate). In the presence of extraneous content, hours of attention are discounted by the content tsunami, $\theta_{tsunami}$, and the self-control filter, β. This filter, chosen by the individual, has a value between 0 and 1.[9] A value of 1 suggests no self-control, while a value of 0 reflects perfect filtering when there is zero reliance on algorithmic guidance and zero fake information is accepted. Then, available cognitive bandwidth hours, which can be interpreted as the supply of attention, can be formulated as:

$$\text{Supply of attention} = h_a * e^{-\beta\theta_{tsunami}}$$

When the demand for attention exceeds the supply, we have an attention deficit. In Fig. 6.3, when $\theta_{tsunami}$ is small, available attention is close to the maximum. As $\theta_{tsunami}$ increases, available bandwidth hours are discounted below the desired hours, h_a, so there is less attention available for either good or bad content. With perfect self-control, $\beta = 0$, and there is no attention deficit. If there is some self-control so that $\beta = 0.33$, then we have less attention available and the deficit increases. As self-control decreases, β increases to 0.66 and the deficit widens. Similarly, when the tsunami increases from level 3 to level 5, the deficit grows, while a tsunami at level 0 signifies maximum cognitive bandwidth hours, which is exactly equal to the rationally chosen hours h_a.

This diagram summarizes personal choices for self-control and desired hours of content consumption: β is a filter representing self-control, $\theta_{tsunami}$ is the hours devoted to consumption of content tsunami and attention deficit is the difference between *demand for cognitive bandwidth* hours and *supply of available* hours of attention. Individuals with perfect filtering will face no attention deficit however large their desired content consumption. Similarly, persons with no desire for content, perhaps due to isolation, also exhibit no attention deficit. On the other hand, a

---

[9] Self-control is ego-depleting, so it works as a drain on self-esteem. Athletes often talk of flow, which is effectively concentration without needing control. But ego depletion can lead to tension. "People who are cognitively busy are also more likely to make selfish choices, use sexist language, and make superficial judgments in social situations" (Handel and Schwartzstein 2018, p. 41).

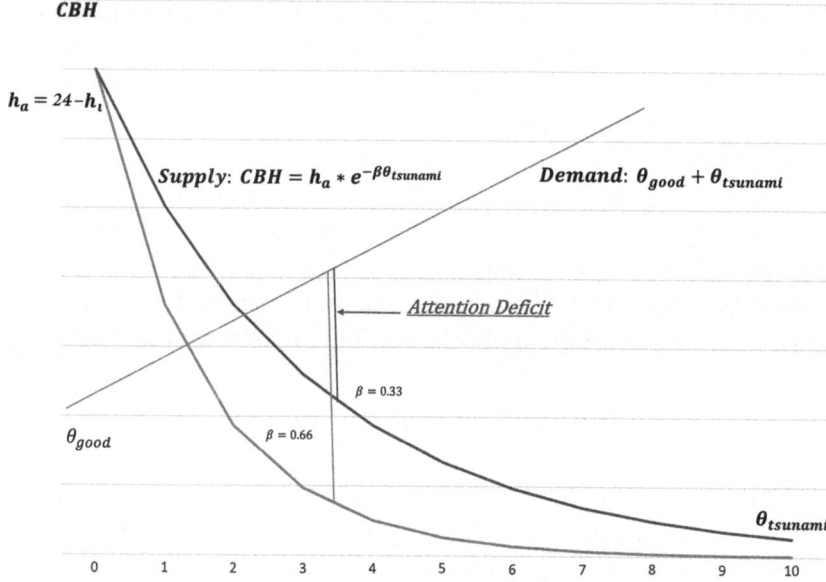

Fig. 6.3 Attention deficit in cognitive bandwidth hours (CBH). Attention Deficit = Demand for CBH minus Supply of CBH. (Source: Author's calculations. When content tsunami is zero, $\theta_{tsunami} = 0$, and all available cognitive bandwidth hours can be utilized for consumption of good content, θ_{good})

high desire for content combined with weak filtering capacity and greater algorithmic outsourcing will increase the attention deficit.

Students of microeconomics will find this diagram perplexing since demand curves usually have a downward slope, while supply curves are upward sloping. This is more of a semantic issue since both demand and supply are from the perspective of the same individual: while one may want to consume more content (represented by the upward-sloping demand for cognitive bandwidth hours), one can only consume according to one's processing capacity and the limits of time (or supply of cognitive bandwidth hours). The representation in the current diagram clearly illustrates the attention deficit.

Chapter 7 explains how cognitive bandwidth shortage reduces the capacity for making intertemporal choices and thinking about the future, which represents attitudes toward uncertainty.

References

Acemoglu, Daron and Pascual Restrepo. 2018. *Automation and New Tasks: The Implications of the Task Content of Production for Labor Demand*. Working Paper, November 2018 and forthcoming in *The Journal of Economic Perspectives*.

Akerlof, G. and R. Kranton. 2005. Identity and the Economics of Organizations. *Journal of Economic Perspectives* 19 (1).

American tech giants are making life tough for startups. *Economist*, June 2018. Accessed 4/5/2019 from https://www.economist.com/business/2018/06/02/american-tech-giants-are-making-life-tough-for-startups

Baumeister, Roy E., Ellen Bratslavsky, Mark Muraven, and Dianne M. Tice. n.d. *Ego Depletion: Is the Active Self a Limited Resource?* Accessed 3/12/18 from https://faculty.washington.edu/jdb/345/345%20Articles/Baumeister%20et%20al.%20(1998).pdf

Bhatt, S. 2017. *How Digital Technology Shapes Markets*. New York: Palgrave Macmillan.

Box Office Mojo. Accessed 5/10/2019 from https://www.boxofficemojo.com/alltime/

Comey in Congressional Testimony. Accessed 5/5/2019 from https://www.c-span.org/video/?401606-1/fbi-director-james-comey-oversight-hearing-testimony

Foer, Franklin, 2018. "*The Era of Fake Video Begins*", https://www.theatlantic.com/magazine/archive/2018/05/realitys-end/556877/?utm_source=eb

Gordon, Robert J. 2016. *The Rise and Fall of American Growth: The U.S. Standard of Living since the Civil War*. Princeton: Princeton University Press.

Handel, Benjamin and Joshua Schwartzstein. 2018. Frictions or Mental Gaps: What's Behind the Information We (Don't Use and When Do We Care? *Journal of Economic Perspectives*, Winter, 32(1).

James, William. 2019. *Talks to Teachers*. Accessed 3/22/2019 from https://www.uky.edu/~eushe2/Pajares/ttpreface.html

Kahneman, Daniel. 2011. *Thinking Fast and Slow*. New York: Farrar, Strauss & Giroux.

Liu, Ernest, Atif Mian and Amir Sufi. January 2019. *Low Interest Rates, Market Power, and Productivity Growth*. NBER Working Paper 25505.

Martin, Katherine Riley et al. 2018. "*Real World Observations Using Virtual Reality Treatments for Anxiety and Related Disorders*" 1235, http://urn.kb.se/resolve?urn=urn:nbn:se:su:diva-155049

Microsoft. 2019. https://docs.microsoft.com/en-us/windows/desktop/debug/system-error-codes%2D%2D0-499-

Mullainathan, Sendhil and Eldar Shafir. 2013. *Scarcity: Why Having Too Little Means So Much*. New York: Times Books, Henry Holt and Company, LLC.

Newport, Cal. 2019. *Digital Minimalism: Choosing a Focused Life in a Noisy World*. Portfolio Publishers.

NYT, *The Privacy Project*, April 10, 2019. Accessed 4/12/19 from https://www.nytimes.com/2019/04/10/opinion/sunday/privacy-capitalism.html?action=click&module=Opinion&pgtype=Homepage]

Pollay, Richard. 1986. *Journal of Marketing*. April 1986

Privacy Rights Clearing House, 2019. *Data Breaches*. Accessed 5/6/2019 from https://www.privacyrights.org/data-breaches

Rogers, Phoebe. 2019. *Breaching the Perimeter: An Exploration of Patterns in Data Breach Environments 2005–2018*. Senior Thesis submitted to the Department of Economics, Princeton University.

SCOTUS. Accessed 5/6/2017 from https://www.supremecourt.gov/opinions/17pdf/16-402_new_o75q.pdf

Sinclair, Stacey, Andreana C. Kenrick and Drew S. Jacoby-Senghor. 2014. *Whites' Interpersonal Interactions Shape, and Are Shaped by, Implicit Prejudice*. Policy Insights from the Behavioral and Brain Sciences. Vol. 1.

Storhart, C., Mitchum, A. and Yehnert, C. 2015. The attentional cost of receiving a cell phone notification. *Journal of Experimental Psychology: Human Perception and Performance*. Accessed 3/22/2019 from https://www.ncbi.nlm.nih.gov/pubmed/26121498?mod=article_inline#

United Nations Human Rights. 2019. *The Right to Privacy in the Digital Age*. Accessed on 7/14/2019 from https://www.ohchr.org/en/issues/digitalage/pages/digitalageindex.aspx

US Department of Health and Human Services, Office for Civil Rights. Accessed 5/6/2019 from https://ocrportal.hhs.gov/ocr/breach/breach_report.jsf

Verghese, Abraham, 2016. *How Tech Can turn doctors into clerical workers*. Accessed 5/19/2018 from https://www.nytimes.com/interactive/2018/05/16/magazine/health-issue-what-we-lose-with-data-driven-medicine.html

Waldrop, Mitchell. 2017. *Virtual Reality Therapy Set for a Real Renaissance*. http://www.pnas.org/content/pnas/114/39/10295.full.pdf

Zuboff, Shoshana. 2019. *The Age of Surveillance Capitalism: The Fight for a Human Future at the New Frontier of Power*. New York: Hachette.

7

Diminished Risk-Taking

Does attention scarcity diminish risk-taking or boldness in everyday life and in business? We saw in the simple model in Chap. 6 that content overload creates an attention deficit. Reduced cognitive bandwidth inhibits the capacity for making sound judgments and clear decisions. As a result, the safe bet is to follow simple rules of thumb or heuristics that generate automatic responses and require little mental effort. A mental shortcut would be to follow the traditional path, well traveled over the past decades. Risk-taking involves experimentation, out-of-the-box thinking, venturing into new territory and imagining unknown possibilities.[1] And it exacts cognitive bandwidth hours of attention.

Cognition and Risk-Taking

In general, there are two different exploratory styles. One style leans toward guidance provided by others—an assisted approach, while the independent thinker's style suggests a high willingness to explore through trial and

---

[1] Indulgence in behavior that is injurious to self or others is not risk-taking when the probability of harm is well documented and close to 100 percent—it is foolhardy.

© The Author(s) 2019
S. Bhatt, *The Attention Deficit*, https://doi.org/10.1007/978-3-030-21848-5_7

error—the engineering and computer scientists approach as exemplified in Mark Zuckerberg's famous motto "Move fast and break things." Whether people try new endeavors, and the success with which these endeavors are met, depends upon their learning strategy. If all new ventures have engineering-like solutions, you have to ask what percentage of the general population is trained to think like engineers? On the other hand, in the assisted approach, guidance provided by the final users may lead to improvement—a more efficient way to solve an old problem—and innovation, which is a new solution to an old problem. But it may not lead to invention, which is an altogether new, untapped, functionality.

Improvement and innovation proceed incrementally. Regardless of the approach, they circle back to the same basic idea. Perhaps another fitness app for the smartphone, with marginal adjustments. Increasingly sophisticated solutions searching for a perfectly matching problem. Rules of thumb have evolved to simplify mental processes. But when applied to innovation and entrepreneurship, this manifests as risk aversion. New ideas rarely arise by applying shortcuts based on past ventures. The iPod was not similar to any past device, nor was the telephone; the smartphone added the first two functionalities but invented the combination of data connectivity, computation and voice. Invention occurs when new problems, previously shrouded in ambiguity, are clearly envisioned and addressed by the entrepreneur. Tinkering, for instance, is like cutting cloth by imagining an entirely new garment instead of applying previously used patterns.

Behavioral economics posits the frequent use of heuristics or the substitution of an easier question for a more complex one to minimize cognitive effort. Under normal conditions individuals prefer situations of cognitive ease (see discussion in Chap. 4). But when people are stressed under a content tsunami, these heuristics take on an added charge. We use *judgment* heuristics when translating questions across dimensions to ease mental effort, effectively substituting a simple question for a harder one. This is the logic behind inferring above-average intelligence from above-average height. Basically, we simplify the problem so that decision-making requires fewer steps.

Two noteworthy heuristics are the availability heuristic and the associative heuristic (Kahneman 2011). Under the *availability* heuristic, when more instances of the current question come to mind, frequency becomes the relevant decision criteria. If you are asked how risk averse you are on a scale of 1–5, you are more likely to say 5 if you have recently written checks for home and auto insurance and purchased extended product warranties. In fact, this recollection might further incentivize you to purchase bicycle insurance.

Risk is not objective but depends upon the metric (Kahneman 2011, p. 141). The availability cascade is an expanded notion of availability heuristic. Availability provides a heuristic beyond frequency of past occurrence; the importance of an idea is often judged by the fluency and emotional charge with which that idea comes to mind. An availability cascade is a self-sustaining chain of events, which may start from media reports of a relatively minor event and lead up to public panic and large-scale government action.

On some occasions a media story about a danger catches the attention of a segment of the public, which becomes aroused and worried. This emotional reaction becomes a story in itself, prompting additional coverage in the media, which in turn produces greater concern and involvement. The danger is increasingly exaggerated as the media compete for attention-grabbing headlines. The response of the political system and business establishment is guided by the intensity of public sentiment. This leads to what Cass Sunstein calls availability cascades, defined as public concern or media coverage as a fraction of the true probability of harm (Sunstein 2017). Media hype increases the former, while the latter is calculated by scientists. A content tsunami covering a recent hurricane might trigger the availability cascade when public concern increases in a cacophony of noise rather than truth. Or these cascades could vanish, as when the public, unthinkingly, ignores media concern, as is the case with global warning and climate change.

The associative mind traces out a coherent pattern of ideas from memory, so the *associative* heuristic leads to stereotyping or representativeness. You are more likely to install a home security system if you observed similar behavior from your neighbors in a housing development. These decision heuristics or mental shortcuts do not imply risk aversion on your part. It is simplifying your decision-making.

Decision-making under conditions of uncertainty requires two tools: better prediction about possible states of the world and judgment about which states are actually relevant. Impaired judgment can therefore have deleterious effects even with massive data processing and data collection technology. AI has allowed data scientists to manipulate this data for superior prediction. However, better prediction is not enough; we need human judgment. Agrawal et al. show that if a particular state of the world is unlikely to manifest, then there is little benefit to understanding the outcome in that state, but judgment is required to eliminate this state from the decision process. "Judgment is exercised when the objective function for a set of decisions cannot be described (i.e. coded)" (Agrawal et al. 2018).

Data availability does not preclude the need for human judgment. When data is minimal, only human judgment can perceive upside gains and therefore make valuable risky investments, while when data is plentiful, and prediction is precise, individuals are more likely to take risky actions and human judgment is required to properly assess hidden downside risks (Agrawal et al. 2018).

A survey of 885 institutional venture capitalists, conducted between November 2015 and March 2016, found that 17 percent of early-stage investors did not use any financial data analysis. "Few VCs use discounted cash flow or net present value techniques to evaluate their investments" (Gompers et al. 2016). VCs and private equity investors used internal rates of returns and multiples of invested capital. The survey also found that 47 percent of the VCs considered management the most important factor in their investment selections compared with industry, market, product or business model. Clearly, judgment is valued over other business metrics (Gompers et al. 2016).

Both these theoretical and empirical results suggest that cognitive effort in the form of judgment is required for properly assessing risk and making investment decisions. An investment in a groundbreaking innovation, not yet crunched by data, is less likely to be made under conditions of poor judgment. Bold strategies require sound judgment, not simply massive data.

Examining the relationship between cognitive ability and risk preferences using data from observed choices in lotteries as well as self-reported risk preferences in survey data, Dohmen et al. find that "cognitive ability tends to be positively correlated with avoidance of harmful risky situations, but it tends to be negatively correlated with risk aversion in advantageous situations" (Dohmen et al. 2018).[2] In other words, higher ability individuals avoid downside risk but avoid situations which could have upside gains. This result lends support to the theoretical findings of Agrawal et al. if we posit that cognitive ability is a reasonable metric for judgment: making risky investments requires judgment when data are scarce, but people with higher cognitive ability avoid precisely these kinds of risks.

Risk preferences are influenced by the environment. Individuals growing up in an environment that emphasizes organization and planning are more likely to develop the relevant cognitive skills for risk management. Choice of a skill-enhancing kindergarten, for example, is itself a "safety choice," exhibiting preferences toward stability and long-term skill enhancement. "Latent risk preference could also play a role later in life in choices about investment in education, with the latter fostering improved cognitive ability" (Dohmen et al. 2018). Interestingly, studies based on self-reported surveys on willingness to take risk suggest that low cognitive ability individuals perceive themselves as relatively risk averse compared with high-ability individuals (Dohmen et al. 2018).

Generally, risk-taking and entrepreneurship can be measured by the likelihood of making a risky investment or founding a startup. Risk evaluation, influenced by culture and the environment, involves combining data with pattern recognition (cognition), which could be as simple as intuition or more complex requiring higher-order mental cognition. The likelihood of making a risky investment, therefore, is a function of the environment, available data, judgment or cognitive ability.

What is it that constitutes a startup culture? What are the factors that enter into the creation of business dynamism—the process whereby older firms, using stale technologies, exit and are replaced by younger firms? Could it be that a content tsunami and the subsequent impaired judgment negate a facilitating environment and massive data?

Consider the entrepreneurship and demographic data that is consistent with this premise. Individuals who experienced the brunt of the

---

[2] The authors note that studies of individual risk preference assume that (1) individuals understand probabilities and that there are no mental gaps or frictions like search costs; (2) other factors, such as wealth, can be controlled for and (3) stated preferences are immutable to the framing of the risky choice.

wealth destruction caused by the collapse in both house prices and the stock market, Gen X, were likely to be 28–43 years old in 2008 (see Fig. 7.6). If the principal founders of businesses are in this cohort, then, overhung with memories of the recent past and their wealth loss, they are likely to be averse to founding new firms. It is the younger cohort, the millennials, who would have been 12–27 years old in 2008 and are recently entering the workforce that could drive future growth by creating new businesses. In 2019 millennials, now 23–38 year of age, are investing in their future by buying stocks, via employer-supported pension funds, instead of creating new jobs.

Admittedly, this demographic experienced the very real burden of constantly striving amidst a swiftly moving economic and political landscape. As discussed in Chap. 1, this was also the generation that was carefully protected or hothoused by over-parenting. The parental urge to protect went into overdrive, which meant that risk experiences were curtailed and that these young adults had no practice at failure and resilience. Responsibility was outsourced so as to avoid making mistakes. Therefore, is it surprising that this very generation is today outsourcing entrepreneurship to firms already listed on the stock market. The stock market is a known risk with plenty of data to support prediction. New ventures on the other hand are an unimagined risk. Cognitive apathy, created by information overload, generates reluctance to undertake these risky activities.

Evidence from National Entrepreneurship and Demographic Data

Careful appraisal of signals from the data and the environment are necessary to make informed risk choices. Consider the fact that stocks started their decade-long upswing on March 9, 2009. Simultaneously, however, the US business sector has seen a startup deficit in the past three decades. How can we reconcile these seeming contradictions? The judgment to plow money into stocks, while simultaneously exhibiting sluggishness with respect to creating new firms, new tasks, is a confounding paradox of the digital economy. While it may be true that the risk profile of a particular stock is in safer territory than that of a startup, creative financing can limit the financial exposure for the founder of a young firm. In fact, the risk exposure for individuals working at firm X to be paid in terms of the stock of firm X is extremely high. If firm X goes bankrupt, not only do you lose your salary and benefits, but you also lose retirement savings.

Figure 7.1 shows that the current period, 2009–2019, is the second longest running bull market: March 2009 was the turning point when the S&P 500 was 57 percent lower than its 2007 peak. Home prices, most households' greatest source of wealth, were collapsing through 2012. The S&P/Case-Shiller Home Price Index declined 25.81 percent between January 2006 and January 2012, while median household income in the US dropped by 7.11 percent over the same period. The upswing in stock prices started well before the upturn in home prices or household income (WRDS and FRED 2019). How is it that the stock market has such an optimistic vision of growth and innovation for the economy, while the data for new businesses and productivity, as shown in Fig. 7.2, suggest otherwise?

Are good ideas becoming elusive? Is it more difficult to innovate today than three decades ago when information and communications technology was nascent and the World Wide Web only a few months old? Using data on intellectual property products (see Fig. 1.1), Bloom et al. find that "research productivity for the aggregate U.S. economy has declined by a factor of 41 since the 1930s, an average decrease of more than 5% per year" (WRDS and FRED 2019). If economic growth is the product of research productivity and research effort, measured by number of researchers or research expenditures (intellectual property products), then steady growth occurs despite falling research productivity simply because of the increase in research expenditures. In the semiconductor industry, the search for a general-purpose technology has perhaps led to an exponential growth in the number of researchers, but "At least as far as semiconductors are concerned, ideas are getting harder and harder to find." Research productivity is declining at the firm level, as measured by the increase in the number of researchers required to sustain the same level of revenue growth.[3] Fifteen times more researchers are required today compared with 30 years ago, to produce the same revenue growth rate (WRDS and FRED 2019).

---

[3] The authors examine 15,128 publicly listed firms on Compustat from 1980 to 2015. Research productivity declined by about 9 percent per year, averaging across the entire sample (WRDS and FRED 2019).

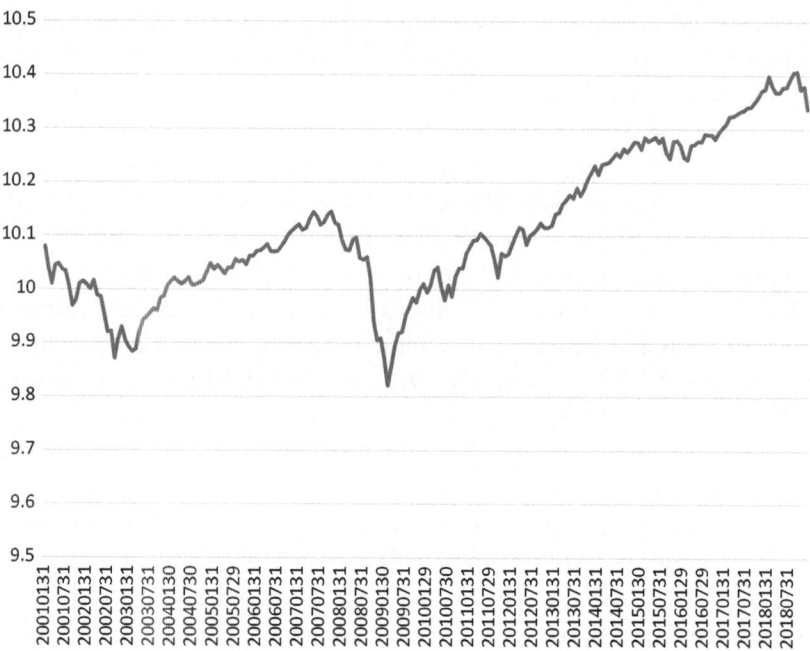

Time Period	Real Median Household Income	Case-Shiller Home Price Index
2006-01-01	$58,746	180.83
2012-01-01	$54,569	134.166
Percent change 2006−2012	−7.11%	−25.81%
2017-01-01	$61,372	184.795
Percent change 2012−2017	12.47%	37.74%

Fig. 7.1 S&P 500 Data: 2001–May 10, 2019 (logarithmic scale); Real Median Household Income and Case-Shiller Home Price Index 2006, 2012 & 2017. (Source: Federal Reserve Bank of St Louis Economic Research. Retrieved on 6/5/2019 from https://fred.stlouisfed.org/series/MEHOINUSA672N [for household income] and from https://fred.stlouisfed.org/series/SPCS20RSA [for Case-Shiller data]; and from Wharton Research Data Services [WRDS], retrieved on 6/5/2019 from https://wrds-sol2.wharton.upenn.edu/output/5e991c6ce39a13da.html#   and   author's calculations)

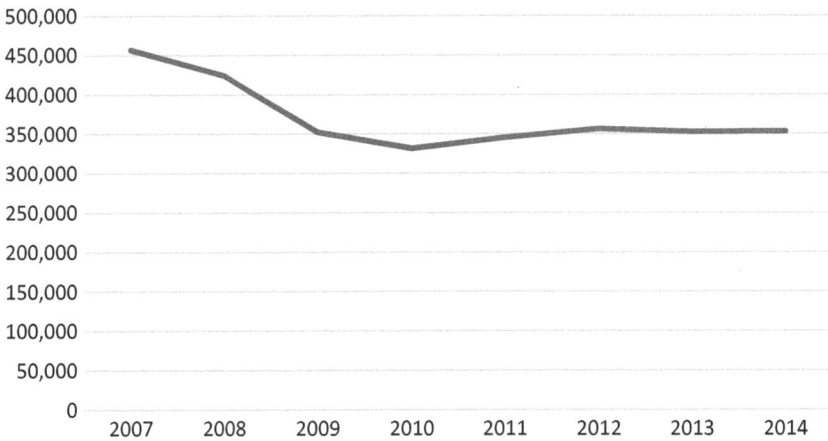

Fig. 7.2 Number of startups. (Note: Number of firms less than one year in existence and with 1–4 employees. Data source: Author's calculations based on data from https://www.census.gov/ces/dataproducts/bds/data.html)

The US business sector has experienced a startup deficit in the past three decades, as shown by Fig. 7.3, reproduced from Alon et al. (2018). The data are from the US Census Bureau, covering the non-farm business sector between 1980 and 2012. This suggests that the business sector is aging, impeding the dynamism that would arise when there is firm entry and exit with an associated reallocation of workers to young, high-productivity sectors from older, lower-productivity sectors (Bloom et al. 2017; Alon et al. 2018).

Reinforcing this picture, Fig. 7.4, reproduced from the Kaufman Foundation Report, shows that the rate of new entrepreneurs, the fraction of US non-business owner adults who start a new business each month, has remained within a band of 0.30–0.34 percent since 2007.[4] It rose in 2009 and 2010 to 0.34 percent and has settled at 0.31–0.33

---

[4] According to the Kaufmann Report, the rate of new entrepreneurs captures the percentage of the adult, non-business owner population that starts a business each month. This indicator captures all new business owners, including those who own incorporated or unincorporated businesses, and those who are employers or non-employers. The rate of new entrepreneurs is calculated from a special panel dataset created from the Current Population Survey (CPS), a monthly survey conducted by the US Bureau of the Census and the Bureau of Labor Statistics.

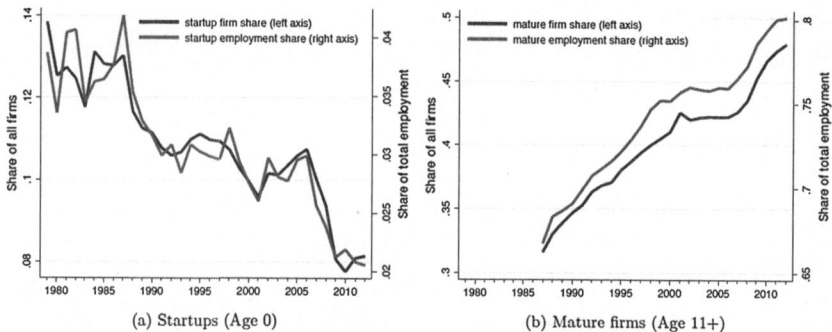

(a) Startups (Age 0)   (b) Mature firms (Age 11+)

Fig. 7.3 Startup firm share and employment share. (Note: Data from US Census Bureau Business Dynamic Statistics. Startup (mature) firm share is the number of age 0 (age 11+) firms as a share of total firms, and startup employment share is the annual employment in age 0 (age 11+) firms as a share of total payroll employment. Reproduced from Alon, Titan M., David Berger, Rob Dent, and Benjamin Pugsley, 2018. Older and Slower: The Startup Deficit's Lasting Effects on Aggregate Productivity Growth. *Journal of Monetary Economics 93*, January 2018, https://ars.els-cdn.com/content/image/1-s2.0-S0304393217301113-gr1_lrg.jpg)

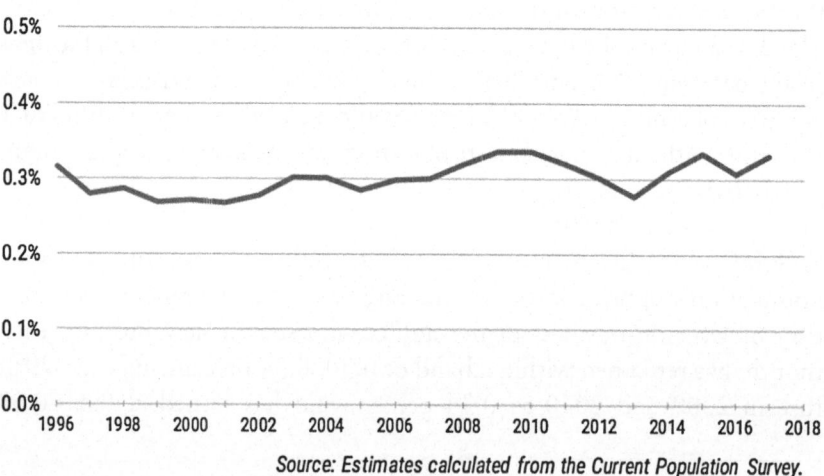

Source: Estimates calculated from the Current Population Survey.

Fig. 7.4 Rate of new entrepreneurs (1996–2017). (Source: Reproduced from the Kaufmann Foundation Report)

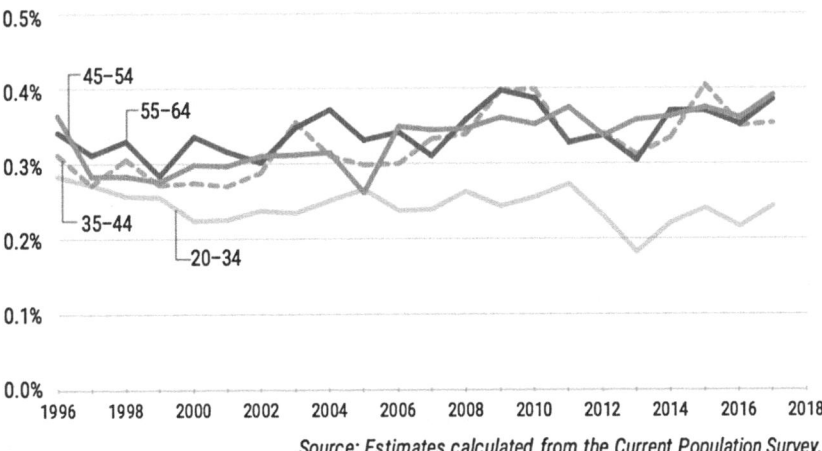

Fig. 7.5 Rate of new entrepreneurs by age (1996–2017). (Source: Reproduced from the Kaufmann Foundation Report)

percent over 2014–2017. If the introduction of the smartphone in 2007 heralded the promise of a foundational tool for mobile peer-to-peer business, then its widespread impact is yet to be felt.

Notably, the rate in 2016 was the lowest among the 20–34 age group, at 0.24 percent, and highest for the 45–54 group, at 0.39 percent, as shown in Fig. 7.5. The millennial generation is shown to be the weakest at starting new jobs, while Gen X was the most entrepreneurial.

The Pew Research Center defines the generation, according to Fig. 7.6, as follows: millennial refers to the generation born between 1981 and 1996, so they would be 22–37 in 2018. Gen X, those born between 1965 and 1980 would be 38–53 in 2018 and the boomers, those born between 1948 and 1964, and would be 54–70. The most entrepreneurial demographic in Fig. 7.5 is the 45–54 year-old group and the 55–64 year-old group or Gen X and the younger boomers (Decker et al. 2017). The weakest are the millennials.

The early startup survival rate is the percentage of new firms that are still active after one year. As Fig. 7.7 shows, this indicator is no higher in 2017 than it was a decade earlier, in 2006.

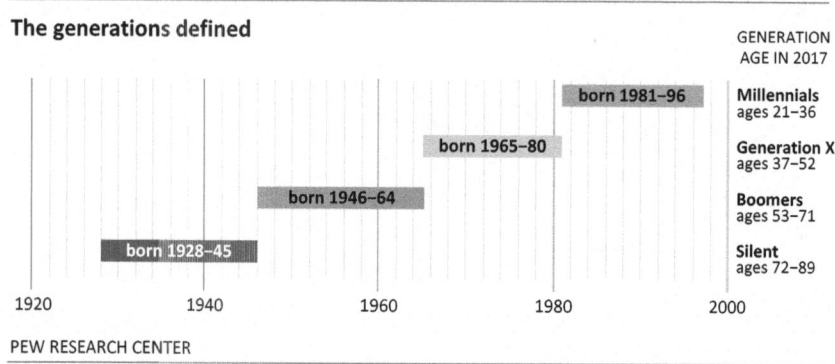

Fig. 7.6 The generations defined. (Source: http://pewrsr.ch/2Dys8lr)

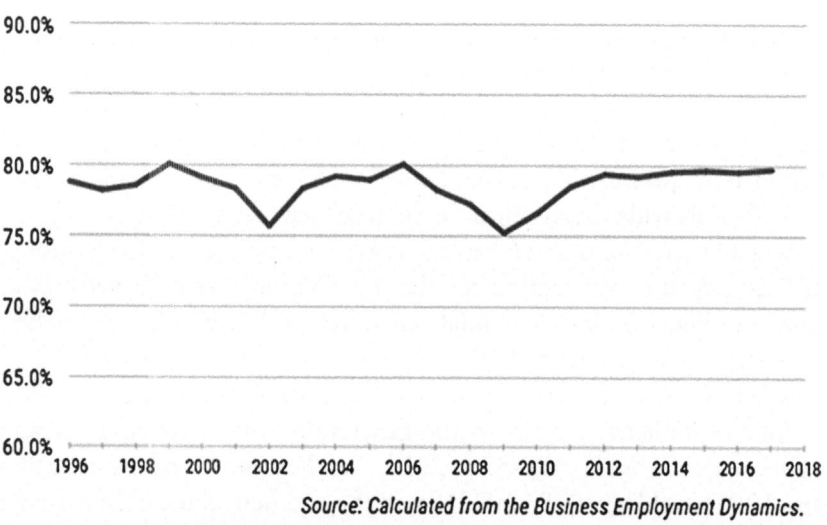

Source: Calculated from the Business Employment Dynamics.

Fig. 7.7 Early startup survival rate (1996–2017). (Source: Reproduced from the Kaufman Foundation Report)

A more recent picture is provided by following the investment funding trail.

There are roughly five stages of venture capital financing: Angel investing or pre-seed funding that usually consists of family and friends funding and crowdfunding. The second stage is the seed stage followed by the later stage comprising series A and series B funding for more established firms. There is also some pre-IPO funding which occurs when the firm is well established in terms of revenue (Decker et al. 2018).

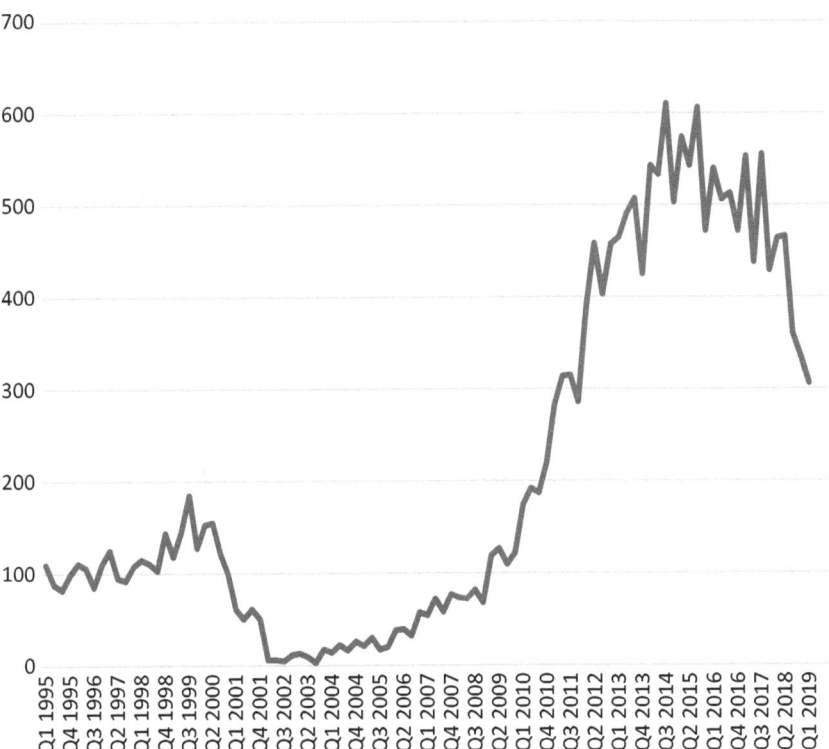

Fig. 7.8 Seed-stage venture capital funding by number of deals; investments and funds in the US. (Source: Author's calculations based on data from PwC/CB Money Insights. Accessed 5/7/2019 from https://www.pwc.com/us/en/industries/technology/moneytree/explorer.html#/currentQ=Q3%202018&qRangeStart=Q3%20 2013&qRangeE)

Defining the staged funding ratio (SFR) as the number of deals in seed stage and series A divided by the total number of deals in later stage, a recent study found that the average SFR grew at a rate of 2.95 percent per quarter over 2002–2018, reaching a peak of 2.36 in early 2016 (Parker et al. 2019). The flow of funds has skewed away from startups, or seed-stage firms since its peak in early 2016 (Fig. 7.8), shifting toward later-stage firms as shown by the declining ratio of seed deals over later-stage deals in Fig. 7.9. Seed-stage funding has been declining since 2016, suggesting a weak prognosis for very young enterprises. Later-stage firms, typically valued at over $1 billion, are receiving relatively more financial resources (Parker et al. 2019).

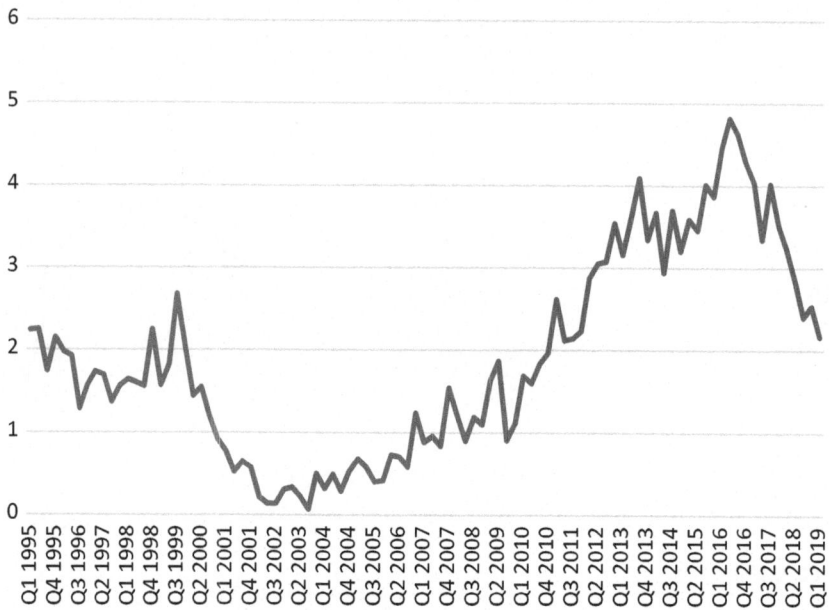

Fig. 7.9 Ratio of number of seed-stage deals over later-stage deals; investments and funds in the US. (Source: Author's calculations based on data from PwC/CB Money Insights. Accessed 5/7/2019 from https://www.pwc.com/us/en/industries/technology/moneytree/explorer.html#/currentQ=Q3%202018&qRangeStart=Q3%20 2013&qRangeE)

The shift in funding shown in Figs. 7.8 and 7.9 signifies a preference for investments with less risk and less variability in payoffs. Diminished risk-taking outcomes such as a declining SFR may be driven by the composition of the limited partners changing from wealthy individuals and foundations to pension funds, who prefer stable cash flows. While crowdfunding may have substituted for conventional venture funding, it operates on a smaller scale.

The data for startup job creation is more dismal.[5] As Fig. 7.10 shows the total number of jobs created by startups in their first year of existence has declined, a variable that can be interpreted as expectation of future business growth.

---

[5] The startup early survival rate, an early-stage indicator of business performance, measures the percentage of new employer establishments that are still active after one year of operation. This indicator is an annual measure calculated from the Business Employment Dynamics (BED).

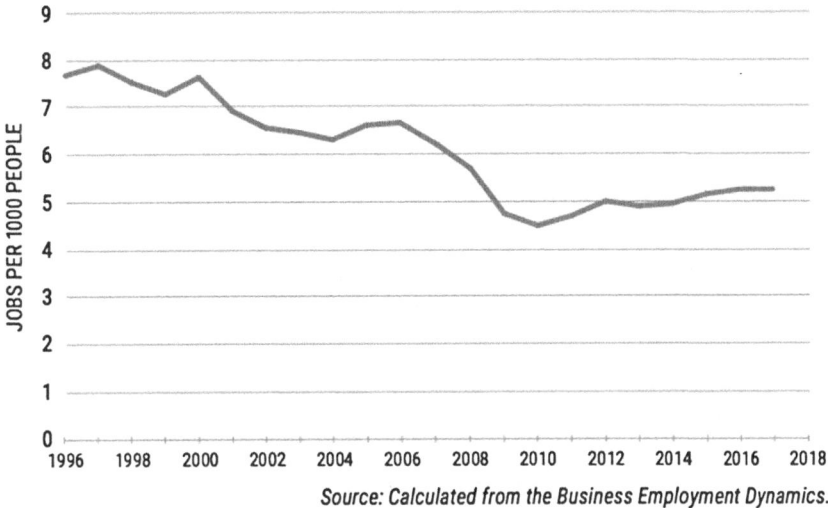

Source: Calculated from the Business Employment Dynamics.

Fig. 7.10 Startup early job creation (1996–2017). (Source: Reproduced from the Kaufman Foundation Report)

Considering labor market data, productivity measured in output per hour has reversed its long-term growth trend. Figure 7.11 shows this dramatic slide over the past decade.

Over the longer period of 1970–2017, average annual rate of labor productivity growth was 1.867 percent. Slicing the numbers from another angle shows that the breakthrough in productivity, overlapping with the introduction of the personal computer in the 1980s and the WWW and search engines in the 1990s, came in 1983–1992 (2.2 percent) and 1993–2003 (2.45 percent). Prior to these years, labor productivity grew by 1.55 percent over 1970–1982, matched by another decline to 1.47 percent over 2004–2017. More precisely, the annual growth rates in recent years of 2015, 2016 and 2017 have been 1.3, 0.1 and 1.1 percent, respectively, which are below the average of 1.47 percent for the period of 2004–2017. Notably, the introduction of the iPod in 2003 and the iPhone in 2007 did not appear to have a significant impact on labor productivity.[6]

---

[6] In fact, the impact of electricity, refrigeration, the internal combustion engine fueled labor productivity growth of *2.82 percent* per year during prior years covering 1920–1970. Gordon calls these the "Great Inventions of the Second Industrial Revolution" (Gordan 2018).

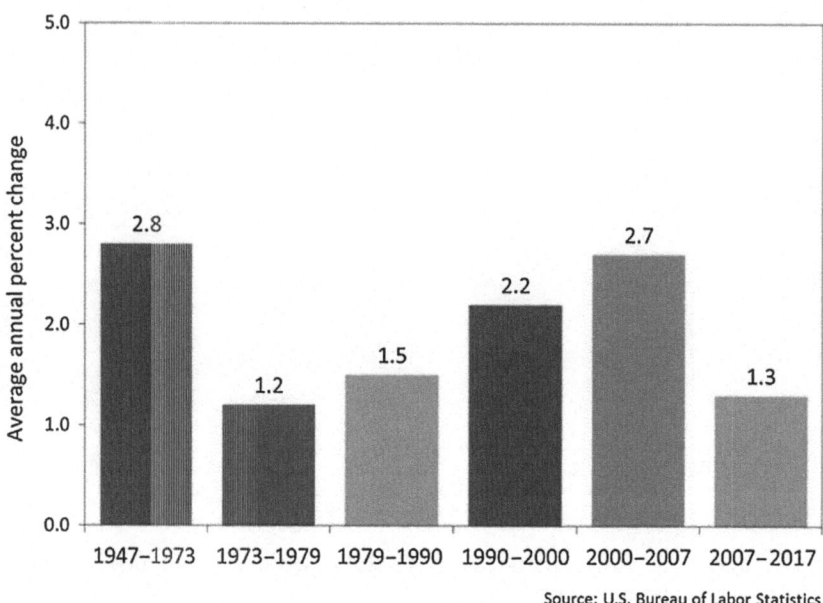

Source: U.S. Bureau of Labor Statistics

Fig. 7.11 Productivity change in the non-farm business sector, 1947–2017. (Source: https://www.bls.gov/lpc/prodybar.htm)

Not only did labor productivity growth slow down, but part-time work has increased while self-employment has shown modest gains after the financial crises. A preference for more flexible work may be the driver, but it could also reflect external stressors motivating workers to readjust their work patterns. For example, the US Labor Department data defines part-time employment as comprising persons who usually work part time for non-economic reasons such as childcare problems, family or personal obligations, school or training, retirement or Social Security limits on earnings, and other reasons. This excludes persons who usually work full time but worked only 1–34 hours during the reference week for reasons such as vacations, holidays, illness and bad weather.

Figure 7.12 shows a steady increase in part-time employment, while Fig. 7.13 shows a steady decline in non-agricultural self-employment (unincorporated workers) from 2002 to 2015 with a modest rise in 2016–2018 (Fig. 7.13). However, as Fig. 7.14 shows, self-employed plus part-time workers, as a percent of total employment, had a somewhat

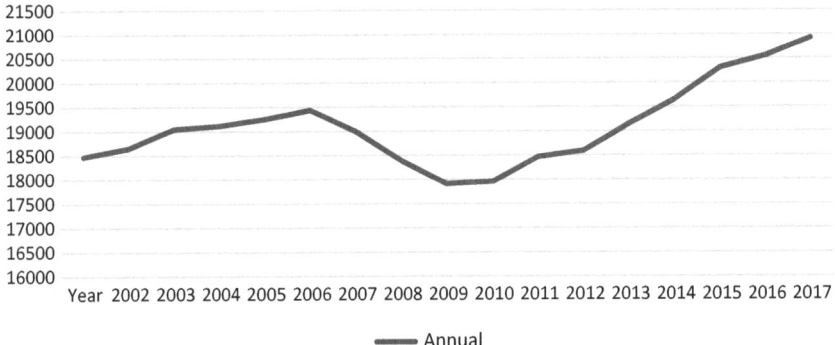

Fig. 7.12 Part-time (for non-economic reasons) non-agricultural workers, in thousands, 16 years and over (seasonally adjusted). (Source: Bureau of Labor Statistics. Accessed on 3/8/2019 from https://data.bls.gov/pdq/SurveyOutputServlet; https://www.bls.gov/cps/lfcharacteristics.htm#self)

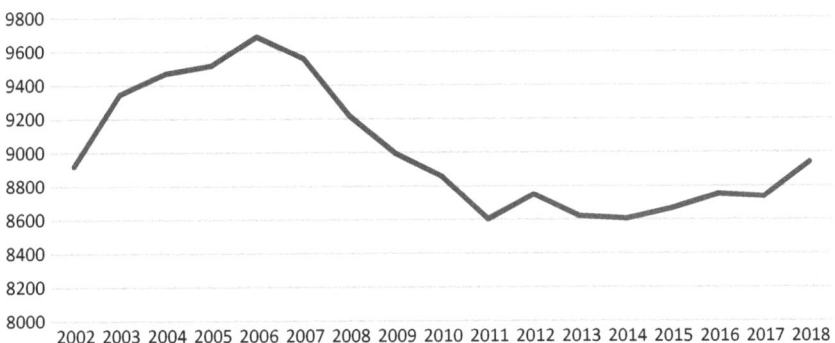

Fig. 7.13 Self-employed non-agricultural workers, in thousands, 16 years and over (seasonally adjusted). (Source: Bureau of Labor Statistics. Accessed on 3/8/2019 from https://data.bls.gov/pdq/SurveyOutputServlet; https://www.bls.gov/cps/lfcharacteristics.htm#self)

muted decline over 2004–2013 with a slow rise from 2014 to 2018, so that gig economy workers declined from 20.2 percent of total employment in 2005 to 19 percent in 2015—a modest *decline of 1 percent*.

While the result in Figs. 7.12, 7.13 and 7.14 are based on national data (Bureau of Labor Statistics), Katz and Krueger have conducted independent surveys with a different conclusion (Katz and Krueger 2016). Much of the difference between the datasets comes from definitions applied to gig

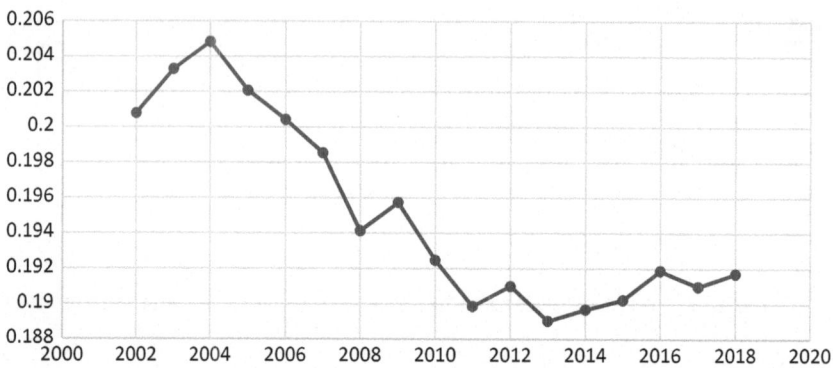

Fig. 7.14 Part-time + self-employed non-agricultural workers as a fraction of total non-agricultural workers, in thousands, 16 years and over (seasonally adjusted). (Source: Bureau of Labor Statistics. Accessed on 3/8/2019 from https://data.bls.gov/pdq/SurveyOutputServlet; https://www.bls.gov/cps/lfcharacteristics.htm#self)

workers. In survey data, for example, there is some confusion and reluctance to clearly state the hours worked. Gig economy workers, defined as those in alternative work arrangements which include independent contractors or freelancers and involuntary part-time workers (temporary help and on-call workers), *rose* from 10.7 percent of the workforce in February 2005 to 15.8 percent in late 2015, a *rise of 5.1 percent*.[7] Of these, 8.4 percent were independent contractors, who were the fastest growing category. Older workers were more likely to be self-employed—23.9 percent of those aged 55–74 were self-employed compared with 14.3 percent of those 25.54. Over 20 percent of gig workers were in the health care and education services in 2015. Katz and Krueger find that tax return data, reported by workers who filled out Schedule C (form 1040 to report self-employment work), are also consistent with this upward trend. But they caution that while the percentage of workers in the gig economy rose, this segment of the workforce is small, concentrated among older workers and could be a substitute for traditional employment, especially during the Great Recession over 2007–2009 (see the discussion in Chap. 1) (Katz and Krueger 2016).

---

[7] Katz and Krueger developed their Rand-Princeton Contingent Worker Survey in late 2015, as part of the Rand American Life Panel, since the Bureau of Labor Statistics was unable to complete the BLS-Contingent Work Survey after 2005.

Recall the discussion in Chap. 1 about the lack of a significant rein-statement effect when demand for labor is shrinking not only due to automation but also because of a *"deceleration in the creation of new tasks"* (Acemoglu and Restrepo 2019). While invention is the creation of vastly new technologies such as the transistor, the creation of new tasks relies on innovation, which represents new ways of solving established, well-defined problems. Innovation adapts around the edges. Innovation is the heart of entrepreneurship and reaches beyond improvement, which is merely more efficient ways of solving established problems. Innovation dives deeper into new organizational and behavioral strategies to define problems with precision and solve them with originality.

Mapping US labor productivity and self-employment (another metric for startups) against demographics is revealing. Consider that the decline in productivity and self-employment that commenced in 2007 was rela-tively flat until 2017 and began increasing in 2018. If we define the *potential* entrepreneurial demographic to include prime-age men and women, those who were between 20 and 34 years of age during the years 2007–2017, then this would correspond to the generation born in the 25 years spanning *1973–1997*. Figure 7.6 defines Gen X as those born between 1965 and 1980 and the millennial generation as those born between 1981 and 1996, so there is close correspondence between younger Gen Xers and millennials and this potential entrepreneurial cohort (1973–1997) cohort.

Not only did these generations come of age during the computer pro-cessing and networking revolutions, the oldest in this entrepreneurial cohort was 11 when the first personal computer, the iMac, was intro-duced in 1984. The youngest, born in 1997, would have been 10 when the iPhone was launched in 2007. These are digital natives who grew up learning to write on the keyboard. They came of age during the data access, information processing and networking revolution, with the introduction of the World Wide Web in 1993 and Google in 1999. They are the generation that would most likely be jump-starting their own businesses in 2017, as eager 20–34-year-olds. What is holding them back? As digital natives, perhaps they are over-connected? Ubiquitous connectivity is part of their lives.

It is possible that connectivity is very local and therefore promotes similar thinking as opposed to innovative thinking. Using Facebook data from April 2016, it was found that individuals connected more with those who lived closest to them geographically. Bailey et al. construct a Social Connectedness Index (SCI) which is the number of friendship links between each pair of counties in the US. Only friendship links between Facebook users in the 30 days prior to April 2016 were used. The probability of friendship between any two Facebook users relative to the product of the total number of users in any two geographical areas declines with distance.[8] A 10 percent increase in distance is associated with a 14.8 percent decrease in the number of friendship ties between those counties, but this declines as the distance gets larger. Overall, for the average US county, 55.4 percent of friends live within 50 miles and over 70 percent of friends live within 200 miles (Gordon 2018)![9] To be sure this finding is narrowly restricted to a single observation (one month in 2016 on Facebook).

The point being made here is that despite technological innovations in information and communication possibilities, the economy has not fired up as had been expected. Connectivity and access to information has not unleashed the creation of new jobs via self-employment. Labor force participation rates have been falling over the past decade, from 66.02 percent in 2008 to 62.85 percent in 2018.[10] The smartphone did not change the way we do business in any meaningful way. Instead, connectivity unleashed a content tsunami, leading to scarcity of cognitive bandwidth, which precluded any adaptation and reorganization of business as well as the introduction of newer businesses.

Apathy in the work environment and in reaching out to new sources of information has been matched by diminished innovation in the creative industries. In general, the risk-return trade-off in the motion picture

---

[8] State–state trade flows are positively correlated with Social Connectedness Index (SCI), but this elasticity is less when controlling for geographic distance. Correspondingly, the elasticity of trade flows with respect to distance is mitigated after controlling for SCI (Katz and Krueger 2016).

[9] On the other hand, counties where people have more geographically dispersed networks (i.e. they have a larger share of friends outside the 100-mile radius) have higher incomes, higher education levels, higher social mobility, higher social capital and life expectancy (Katz and Krueger 2016). Social ties may weaken with distance, but individuals with more distant ties are likely to be socio-economically better off.

[10] BLS https://data.bls.gov/timeseries/LNS11300000

industry is extreme, and producers responded by creating a captive market via the acquisition of theater chains and video stores. Before the age of blockbusters like Steven Spielberg's *Jaws* in 1975, theatrical release was gradual and spread from urban to suburban areas, depending upon the initial critical reaction by "urban sophisticates." Changes in the urban landscape ushered in a new business strategy with aggressive ad campaigns "designed to trigger audience anticipation and drive a massive Friday-night opening across thousands of screens—critics and snobs be damned." Today we are witnessing the demise of the powerful studio and the talent agent, both replaced by ownership of a scarce resource—intellectual property (IP) (Bailey et al. 2018).

But then came streaming technology, first offered by Netflix in 2007, empowering consumers by decoupling time and content, as described earlier. In response, the movie studios captured ownership of a critical input, intellectual property, and soon product variety was controlled by a narrow set of major studios. IP or intellectual property was a scarce resource as characters in the Marvel Cinematic Universe or the Star Wars or Harry Potter universe were branded. The stars were the characters, not the actors. Uncertainty was eliminated since the story was already written, the comics and books already sold. Controlled variety diminished uncertainty, "[b]ut a minimal standard of human relatability is not being met, on a routine basis, in the medium's dominant genre. People who are nothing like us rescuing a world that is nothing like ours is not a recipe for artistic renewal" (Bailey et al. 2018).

Mental Health and Risk Preferences

Risk aversion can be thought of as the impulse to minimize any unexpectedness, a resistance to change, a reluctance to adapt and, when unanticipated events arise, a desire to preemptively insure against the outcome. In Chap. 1 we saw the alternate forms of uncertainty: risk, where the probabilities of outcomes are known; ambiguity, where the underlying drivers of outcomes and their corresponding probabilities are unknown; and misspecification, where the very context is unknown—the possible outcomes, their causes and the associated probabilities. The current aversion to uncertainty is with respect to all three formulations.

With known probabilities and data, such as detailed actuarial tables used by life insurance companies, the uncertainty can be priced and insurance against outcomes purchased so as to minimize the negative consequences of these eventualities. When the uncertainty becomes more severe, as with ambiguity and misspecification, the risk mitigation becomes risk avoidance. Suppose you are uncertain of the very nature of a weather event—its likelihood, its form and its extent—perhaps because such an event has no record of occurrence, then you may react to unfounded rumors by not leaving your home! This is precisely the response observed in multiple contexts.

Attitudes and proclivities of the teenage population have morphed over the decades in a manner suggesting an increased aversion to risky outcomes. Alcohol, drugs and sex are lower today than a decade ago. Using data from the Centers for Disease Control, a recent report showed teenage participation in drinking was 60.4 percent in 2017, down from 81.6 percent in 1991, while participation in drug use has declined from 24 percent in 1991 to 19.8 percent in 2017 (Youth Anxiety 2019). A new culture of carefully curated lives, helicoptered and snow-plowed by parents, has become pervasive, suggesting diminished risk-taking via ownership of personal decisions and reliance on individual resourcefulness (Acemoglu and Restrepo 2019). Considering that the prime working-age adult in the next decade is going to come from Gen Z, this decline in risky behavior could suggest an overall increase in risk aversion, which bodes ill for risk-taking activity and entrepreneurship moving forward. However, the average stress levels, measured on a scale of 1–10, are 5.3 for this generation, just behind millennials at 5.7[11] (see the discussion in Chap. 1 on anxiety among college students).

A hyper-focus on threatening situations, labeled threat bias, can lead to excessive anxiety. Giving preferential attention to threat, instead of cues for safety, leads downward into a vicious cycle of heightened anxiety, where threat-processing is facilitated and opportunities for minimization of fear reduced (APA 2019). The American Psychological Association conducted a poll in 2018 and found that high-intensity issues such as gun

---

[11] The 2018 Stress in America™ survey was conducted online within the US by The Harris Poll on behalf of the American Psychological Association (APA) between July 27, 2018, and August 28, 2018, among 3458 adults ages 18 and older who reside in the US. Interviews were conducted in English and Spanish.

violence, rise in suicide rates, climate change and global warming, family separations and high-profile sexual abuse cases have dominated content provision and created significant stressors for Gen Z—more so than any other generation. This generation, the youngest in the US, is least likely to report robust mental health, with 45 percent reporting excellent or good mental health, below 56 percent of millennials, 51 percent of Gen Xers and 70 percent of boomers—the lowest among all generations, according to the American Psychological Association (APA) website. Over two-thirds of this generation feel stressed about our nation's future (68 percent) and do not feel that our nation is moving toward being stronger than ever (66 percent) (Pain in the Nation: The Drug, Alcohol and Suicide Crises and the Need for a National Resilience Strategy 2017).

Could loneliness and social isolation be the underlying cause of increased mental illness, anxiety, low productivity and the startup deficit? A 2019 survey by the Edelman Group, a public relations firm, found that 74 percent of the general population and 72 percent of technology executives believed that AI-infused devices will lead to less human interaction and more isolation (Edelman Group 2019; Belkin 2019). Further, 71 percent of the general population and 65 percent of tech executives thought that these AI-infused devices will lead to a "dumbing down of people" so that, according to David Weld, a Professor of Computer Science at the University of Washington, "human abilities may atrophy" (Belkin 2019).

Simultaneously, David Autor writes that the geography of jobs has changed and cities are now the symbol of the disappearance of well-paid middle-class jobs. American migration across state lines fell by 50 percent from 1991 to 2011 (Autor and Salomons 2018). Such movement is at the lowest level since 1948 when the government began to track such figures. Despite the fact that the Internet allows working from a single location, out-of-control housing markets, demand for high-skilled labor and political and cultural divisions induce sluggish mobility. Lack of mobility can exacerbate social and economic divisions.

Despite the uplifting news about entrepreneurial activity for the skilled, educated fraction of the older demographic, the picture is bleaker for others. An international comparison of mortality rates, all ages 50–54, due to deaths of despair, shows that the rates are higher in the US compared with Canada, Australia and Western Europe. A 2018 report by the nonprofit organization, Trust for America's Health and Well Being Trust, shows that over 1 million Americans have died between 2006 and 2015 from drug overdoses, alcohol and suicides, and 44.7 million American adults experienced a mental illness in 2016 (Metcalf 2018).[12] Could this be a result of general anxiety and fear?

Distinguishing between mortality and morbidity, Case and Deaton document that mortality declines for white non-Hispanic men and women during 1998–2014 from cancer and heart disease were offset by significant increases in suicides, drug overdoses and alcohol-related liver mortality, collectively labeled deaths of despair. "Mortality increases for whites in midlife were paralleled by morbidity increases, including deteriorations in self-reported physical and mental health and rising reports of chronic pain" (Katz and Krueger 2016). Their most striking result lies in an international comparison of mortality rates, all ages 50–54, due to deaths of despair. While mortality rates averaged at about 40 per 100,000 after 2000 in France, Canada, Germany and Australia, this number doubled to 80 among US non-Hispanic whites. The average annual percentage change in mortality due to deaths of despair for this age group was a positive 5.4 percent compared with declines in France, Germany, Spain, Italy and Japan. Rates increased by 1 percent in the UK, by 2.5 percent in Canada and Australia, and 3 percent in Ireland.

Mortality rates for non-Hispanic whites without a B.A. increase with age for all cohorts born after 1940, but especially for those born in the cohort after baby-boomers, Gen X. This suggests heavier drinking and abuse of drugs at an increasing rate over successive generations. The correlation between the falling income and rising morbidity rates is coincidental, not causative and so does not appear compelling; long-run social trends are the more likely cause according to Case and Deaton. Pessimism among whites could arise from the knowledge that they are no better off than their parents and that white privilege should be sustained. The long-term decline in wages for those with less than high school education

> began for those leaving high school and entering the labor force after the early 1970s—the peak of working-class wages, and the beginning of the

---

[12] The working-age population in the US, or those between 15 and 65, was 206,322 million or 62.8 percent of total US population (328,327 million in Q4 of 2018), according to data from the Federal Reserve Bank of St Louis.

end of the 'blue-collar aristocracy'—worsened over time, and caused, or at least was accompanied by, other changes in society that made life more difficult for less educated people, not only in their employment opportunities but also in their marriages, and in the lives of and prospects for their children. (Case and Deaton 2017)

Changes in marriage patterns, religious affiliations and career choices left people relatively rudderless, "a loss of the structures that give life a meaning." Could the "collapse of the white working class" provide some insight as to why there is increasing risk aversion and an equivalent collapse in entrepreneurship?

Quite possibly some of the explanation of the substance abuse escalation lies in the misallocation of talent. Note the fact that workers in the category labeled "Substance abuse, behavioral disorder and mental health counselors" in the BLS statistics earn a mean annual wage of $46,560, while the mean annual wage across all occupations is $50,620 (Egan and Dennis-Tiwary 2018). Child, family and school social workers earned a mean annual wage just higher at $48,430, while those who "plan, direct or coordinate the academic and nonacademic activities of preschool and childcare centers or program (excludes pre-school teachers)" earn a mean annual wage of $53,550. Those in the trenches are paid, and hence valued less, than those in administrative jobs.

Availability of massive data encourages fine-tuned data analytics, assigning numbers to events, thereby creating the semblance of certainty.[13] Emotions are exaggerated with numbers, scaling uncertainty with percentages and ratios, seeking assurance by aligning with the majority. When we are told that X percent of adult Americans have faced an episode of depression in their lives, we are more comfortable with our own possible depression. This raises the possibility that existence of data analytics has itself created this aversion to uncertainty. How can we be afraid of the unimagined event? When sailors ventured across the oceans centuries ago, they sought riches or escaped penury—life wasn't so good back

---

[13] It should be noted that not all data analytics provide clarity. In a survey of IT decision-makers and data managers by the data protection firm Veritas, 40 percent of respondents found an overload of management tools and systems among the day-to-day challenges in the management of data, while 38 percent found "too many data sources to make sense of" and 34 percent lacked the skills and technology to harness the power of data (APA 2019).

home so they ventured outward, blithely unaware of what might become of them. If life was comfortable and secure, there would be much to lose if they sailed overseas and probably nothing to gain. So it comes down to where you start from when you consider risky choices.

Our sense of well-being depends upon where we started. The starting point is our point of reference and today our starting point is elevated. Daniel Kahneman (2011) shows the importance of a reference point when considering gains and losses in monetary situations. From a reference point of $0, the gain of $1 is valued less than the loss of $1—the pain incurred with the loss is greater than the happiness with the gain. This effect is accentuated if you start with $100 as compared with a million dollars. Analogously, our reference point consists of many millennium development goals being met: poverty has been wiped out across the globe, disease has been eradicated, severe famine and hunger are on the decline. There is much to be satisfied with and little to pine for except when it comes to gaining more than one's neighbors and friends. This puts individuals in a situation where they don't want to risk their good fortune by taking on risky endeavors and placing themselves in unknown situations.

Elevated levels of living standards together with big data have created a cocoon of safety, and we shudder to venture outside. This is the root of an aversion to taking chances, to encountering risky situations. How does this tie in with my thesis on content tsunami and cognitive apathy? Big data is the proliferation of information, the easy availability of data on all sorts of issues and events, evoking images of a world that we may not like and that is far from the comfort of our current situation. The more we know, the more we become afraid of losing what we already have. Cognitive apathy or attention deficit, however you want to think about the impact of content tsunami, has the fundamental effect of freezing our minds due to information overload. Then we become incapable of thinking clearly, rationally, objectively of the world, and retreat into the safety of the known past. This is the source of nostalgia that is all pervasive today. We are hearing music, seeing movies, in a revival of past fashion, all because the new world is too uncertain, and we know too well how uncertain it is.

Data and a fairly good standard of living combine to create risk aversion. Nostalgia is evidence of risk aversion. Why? Rather than looking ahead, where we see a cloudy future, we prefer to look backward as a reminder of how good we have had it and would like to preserve it as a model for the future.

The 2019 college fraud scandal is an example of people wanting assurance against an uncertain future created merely by the possibility of losing status. These fraudulent payments hijacked an admissions system purported to be objective and merit-based but which has morphed, in recent years, into an admissions jungle crisscrossed by paths labeled athletics, geography, race, trustee and legacy connections. The admissions process has been side-doored by obsessive parents who paid bribes to get their children into reputable colleges, suggesting that parents view elite colleges as a status symbol and networking opportunity, not simply a path to higher income and stable jobs or a sense of personal achievement. Admittedly, the numbers are discouraging: only 2.5 out of every 1000 students attending elite colleges will earn close to $200,000 when their parents' earning are in the bottom quintile (Occupational Employment statistics n.d.).[14] But the parents in the college fraud scandal are in the top 1 percent of the income distribution. Chetty et al. have shown that intergenerational mobility has been flat for recent birth cohorts (Chetty et al. 2014, 2017). For a child born during the years 1971–1993, the chances of climbing into a higher-income distribution category is no higher today than in 1970 as there is much greater likelihood of remaining in the same category—a child's position in the income ladder is closely related to where her parents stand on that ladder. Clearly these parents want a guarantee that this status will be maintained and that their children will acquire the prestige conferred by attendance at elite universities. They would rather break the law than risk their child left behind.

Do universities know this and make exclusivity, a version of prestige, an important part of their product? Universities race to lower admission rates by fanning out across the globe to increase the applicant pool. Harvard accepted 4.6 percent of its applicants in 2018, Princeton accepted 5.5 percent and Stanford 4.3 percent. In 2008, the rates were 8, 9.3 and 9.5 percent, respectively. Parents, meanwhile, race to back-door admissions by hiring college consultants for fees matching those charged

---

[14]Chetty et al. (2017) show that upper tail mobility rates (earning at least $197,000 by age 34, given that parents are from the bottom decile of the income distribution) are 0.25 percent at elite colleges.

by top-tier law firms. The college counseling market "has grown in recent years as anxiety has increased among middle and upper middle-class families about getting their children into a top school" according to one educational consultant (Occupational Employment statistics n.d.).

What are the implications for innovation? Are startups creating new ways of doing things or improving old ways, merely tinkering around with amendments to the existing models? What has Alexa actually achieved? How has it made our lives more meaningful other than saving time or physical effort? Perhaps the innovations are arriving in the form of new services in the media and entertainment world. If we have all we need in terms of material goods, and we have more leisure time, then perhaps we need something to fill this time. Innovation is arising in the provision of new forms of entertainment and the use of our leisure time. The impact may not be equal to that of electricity or the computer revolution, but the soul-searching that is enabled by new services is venturing into new territory of human experience. As we saw in Chap. 6, the entertainment industry is rising to the occasion. In Chap. 8 and the concluding chapter we turn to an examination of broad guidelines for new content.

Appendix: Non-farm Business Sector Annual Growth Rate of Labor Productivity

US Bureau of Labor Statistics

Labor productivity is measured as Output per hour of all persons: output divided by hours.

Output: annual and quarterly indexes of real output are based on chained Fisher-Ideal quantity indexes. Quarterly indexes are adjusted for consistency to the annual indexes.

Hours of all persons: employment times average weekly hours times 52.

Labor input data—hours at work by all persons engaged in a sector—is based on information on employment and average weekly hours collected in the monthly BLS surveys of establishments and households. Weekly hours are adjusted to the hours at work definition using the BLS Hours at Work survey, conducted for this purpose. Data from the National Compensation Survey are used for recent years.

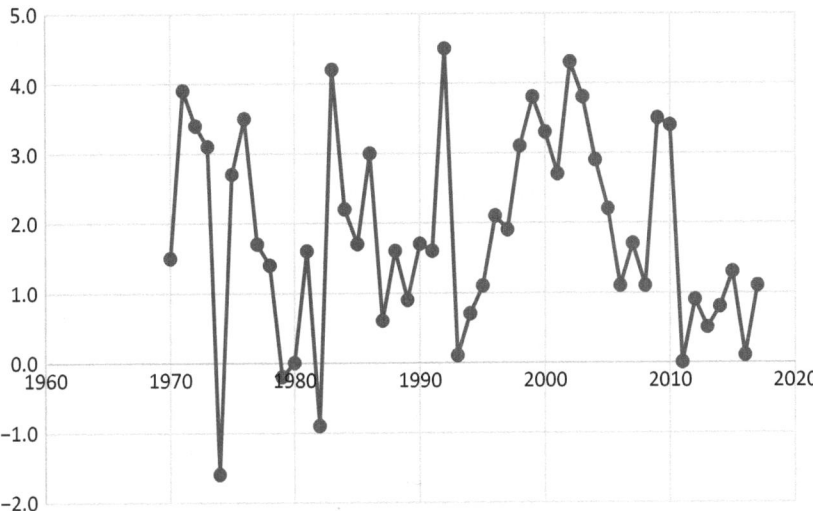

Fig. 7.15 Labor productivity—output per hour of all employed persons 1970–2017. (Source: Author's calculations using data from the Bureau of Labor Statistics. https://www.bls.gov/lpc/#tables)

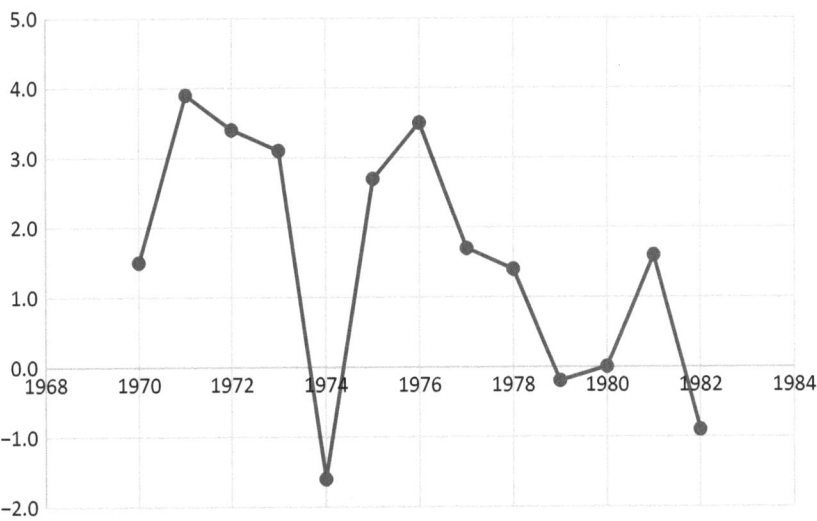

Fig. 7.16 Labor productivity—output per hour of all employed persons 1970–1982. (Source: Author's calculations using data from the Bureau of Labor Statistics. https://www.bls.gov/lpc/#tables)

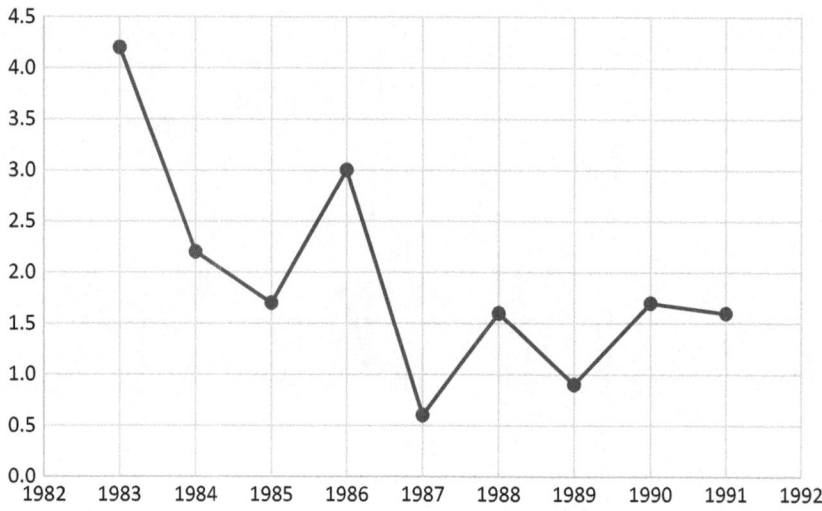

Fig. 7.17 Labor productivity—output per hour of all employed persons 1983–1991. (Source: Author's calculations using data from the Bureau of Labor Statistics. https://www.bls.gov/lpc/#tables)

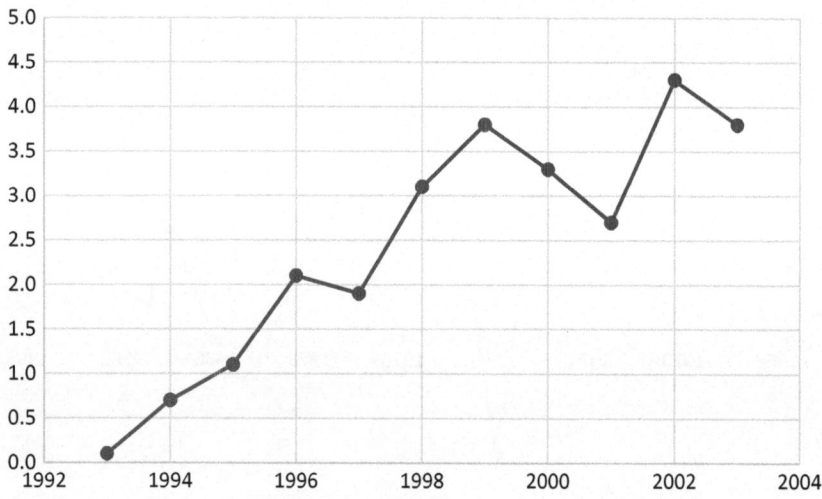

Fig. 7.18 Labor productivity—output per hour of all employed persons 1993–2003. (Source: Author's calculations using data from the Bureau of Labor Statistics. https://www.bls.gov/lpc/#tables)

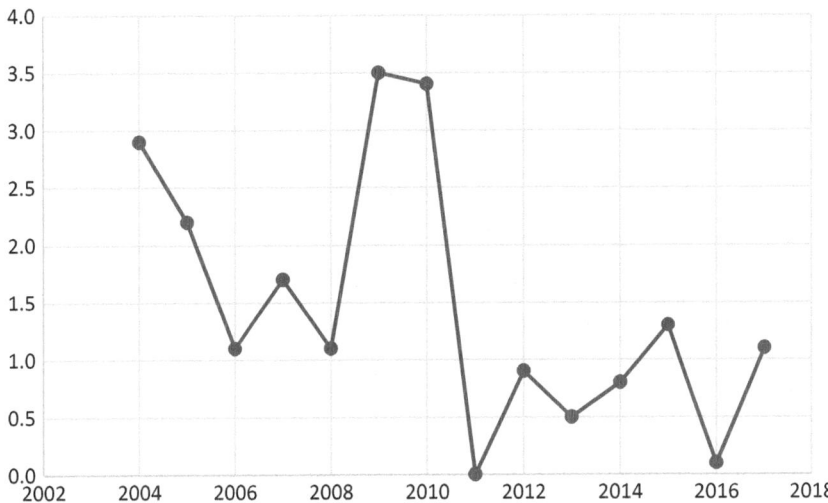

Fig. 7.19 Labor productivity—output per hour of all employed persons 2004–2017. (Source: Author's calculations using data from the Bureau of Labor Statistics. https://www.bls.gov/lpc/#tables)

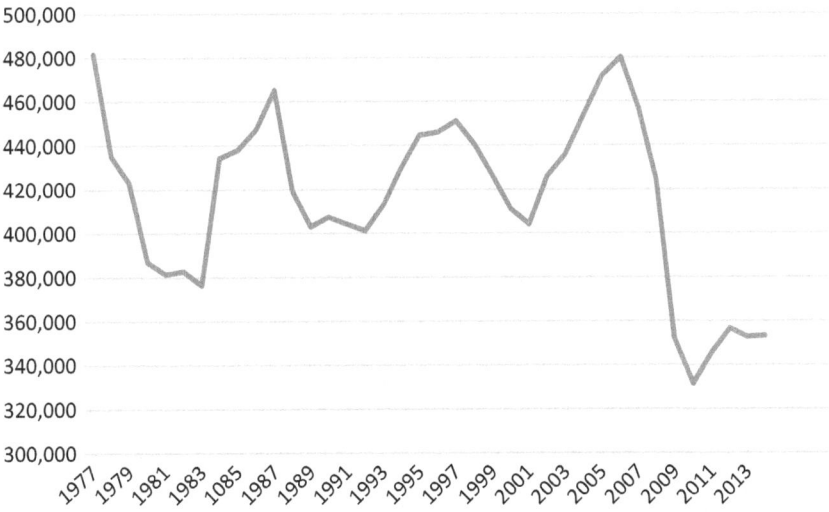

Fig. 7.20 Number of firms with 1–4 employees and less than one year in existence, 1977–2014. (Source: US Census: Business Dynamic Statistics, Longitudinal Business Database, Firm Characteristics Data Tables, 1977–2014, author's calculations. https://www.census.gov/ces/dataproducts/bds/data_firm.html. Unit of Analysis: Active Establishments [Establishment activity is defined by March 12 employment])

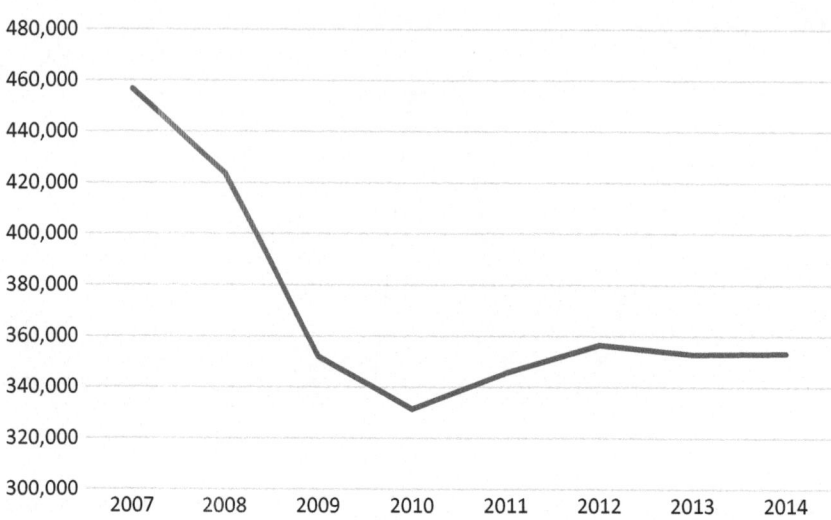

Fig. 7.21 Number of firms with 1–4 employees and less than one year in existence, 2007–2014. (Source: US Census: Business Dynamic Statistics, Longitudinal Business Database, Firm Characteristics Data Tables, 1977–2014, author's calculations. https://www.census.gov/ces/dataproducts/bds/data_firm.html. Unit of Analysis: Active Establishments [Establishment activity is defined by March 12 employment])

References

Acemoglu and Restrepo. 2019. *Automation and New Tasks: How Technology Displaces and Reinstates Labor.* Journal of Economic Perspectives 33(2), 2019.

Agrawal, Ajay, Joshua Gans and Avi Goldfarb. 2018. *Exploring the Impact of Artificial Intelligence: Prediction versus Judgment,* NBER working paper 24626, May.

Alon, Titan M., David Berger, Rob Dent, and Benjamin Pugsley. 2018. Older and Slower: The Startup Deficit's Lasting Effects on Aggregate Productivity Growth. *Journal of Monetary Economics,* January 2018.

APA. 2019. Accessed 3/20/2019 from https://www.apa.org/news/press/releases/stress/2018/stress-gen-z.pdf

Autor, David and Anna Salomons. 2018. Is automation labor-displacing? Productivity growth, employment and the labor share. *Brookings Papers on Economic Activity,* March 8–9, 2018.

Bailey, Michael, Rachel Cao, Theresa Kuchler, Johannes Stroebel and Arlene Wong. 2018. Social Connectedness: Measurement, Determinants, and Effects. *Journal of Economic Perspectives* 32 (3).

Belkin, Douglas. 2019. The Legitimate World of High-End College Admissions, *WSJ* March 13. Accessed 3/14/2019 from https://www.wsj.com/articles/the-legitimate-world-of-high-end-college-admissions-11552506381?emailToke n=05c0baf812d2d73377a0e9fc346c12e71NwQoFQmC3uKm/kXu+gbiVNl/d2SAmKqwF3WZFc5UlKFkCOIhypX7Zi0Q3Syl79NVn-wtNSbGREhpFGFxU6kHCm35wb7m/l8ZWy3VEhCwrL0%3D&reflink=article_email_share

Bloom, Nicholas, Charles Jones, John Van Reenen and Michael Webb. September 2017. *Are Ideas Getting Harder to Find?* NBER Working Paper #23782. (DOI): https://doi.org/10.3386/w23782.

Case, Anne and Sir Angus Deaton. *Mortality and Morbidity in the 21st Century.* March 23, 2017. Brookings Papers on Economic Activity. Accessed 3/4/2019 from https://www.brookings.edu/bpea-articles/mortality-and-morbidity-in-the-21st-century/

Chetty, Raj, John Friedman, Emmanuel Saez, Nicholas Turner, Danny Yagan. 2017. Mobility Report Cards: The Role of Colleges in Intergenerational Mobility. *NBER Paper 23618.*

Chetty, Raj, Nathaniel Hendren, Patrick Kline, Emmanuel Saez and Nicholas Turner. 2014. Is the United States Still a Land of Opportunity? Recent Trends in Intergenerational Mobility? *American Economic Review: Papers and Proceedings* 104 (5).

Decker, Ryan A., John Haltiwanger, Ron S. Jarmin and Javier Miranda. 2017. *Declining Dynamism, Allocative Efficiency, and the Productivity Slowdown,* Working Papers 17–17, Center for Economic Studies, U.S. Census Bureau.

Decker, Ryan, John Haltiwanger, Ron S. Jarmin and Javier Miranda. 2018. *Changing Business Dynamism and Productivity: Shocks vs. Responsiveness,* Finance and Economics Discussion Series 2018-007, Board of Governors of the Federal Reserve System (U.S.).

Dohmen, Thomas, Armin Falk, David Huffman and Uwe Sunde. 2018. *On the Relationship between Cognitive Ability and Risk Preference,* 32 (2), Spring.

Edelman survey, 2019. *Survey of Technology Executives and the General Population Shows Excitement and Curiosity yet Uncertainty and Worries That Artificial Intelligence Could be a Tool of Division.* March 2019. Edelman.

Egan, Laura and Tracy A. Dennis-Tiwary. 2018. Dynamic measure of anxiety-related threat bias: Links to stress reactivity. *Motivation and Emotion*, https://doi.org/10.1007/s11031-018-9674-6

Gompers, Paul, William Gornall, Steven Kaplan and Ilya Strebulaev. September 2016. *How Do Venture Capitalists Make Decisions?* NBER Working Paper 22587. Document Object Identifier (DOI): https://doi.org/10.3386/w22587.

Gordon, Robert. 2018. Declining American Growth Despite Ongoing Innovation. *Explorations in American History*. March 26. Accessed on 5/25/2019 from http://economics.weinberg.northwestern.edu/robert-gordon/files/RescPapers/Declining_growth_innovation.pdf

Kahneman, Daniel. 2011. *Thinking Fast and Slow*. New York: Farrar, Strauss & Giroux.

Katz, Lawrence and Alan Krueger. September 2016. *Alternative Work Arrangements*. NBER #22667.

Median Household Income and Case-Shiller Home Price Index, Retrieved on 5/31/2019 from https://alfred.stlouisfed.org/series?seid=CSUSHPINSA&utm_source=series_page&utm_medium=related_content&utm_term=related_resources&utm_campaign=alfred

Metcalf, Stephen. 2018. How Superheroes Made Movie Stars Expendable, *The New Yorker*, May 28. Accessed 5/23/2018 from https://www.newyorker.com/magazine/2018/05/28/how-superheroes-made-movie-stars-expendable?mbid=nl_Daily%20052118&CNDID=5872155&spMailingID=13553277&spUserID=MTMzMTg1NTU2NjUzS0&spJobID=1401930068&spReportId=MTQwMTkzMDA2OAS2

Occupational Employment statistics. Accessed 4/6/2019 from https://www.bls.gov/oes/2017/may/oes_nat.htm?utm_source=newsletter&utm_medium=email&utm_campaign=newsletter_axiosdeepdives&stream=top#39-0000

Pain in the Nation: The Drug, Alcohol and Suicide Crises and the Need for a National Resilience Strategy, November 2017. Trust for America's Health and Well Being Trust.

Parker, Kim, Nikki Graf and Ruth Igielnik. 2019. Pew Research Center, *Gen Z looks a lot like Millennials on Social and Political Issues*, Pew Research Center, January 17. Accessed 3/2/2019 from http://www.pewsocialtrends.org/2019/01/17/generation-z-looks-a-lot-like-millennials-on-key-social-and-political-issues/

Sunstein, Cass. 2017. *#Republic: Divided Democracy in the Age of Social Media.* Princeton: Princeton University Press.

WRDS and FRED, 2019. Retrieved on 5/31/2019 from https://alfred.stlouisfed. org/series?seid=CSUSHPINSA&utm_source=series_page&utm_ medium=related_content&utm_term=related_resources&utm_ campaign=alfred

Youth anxiety. Accessed Axios on 3/20/19 from https://www.axios.com/drugs-sex-alcohol-losing-appeal-american-teenager-gen-z-d6eef-7cf-369c-46e0-89fb-4a8ead55e728.html

8

Restoring Boldness and Reducing Apathy

When small business startups, firms less than one year in existence, are the main drivers of new products and services, a startup deficit diminishes economic dynamism. In recent years the proximate causes include a content tsunami and cognitive apathy. The decoupling of time and content has allowed asynchrony in communication—when, where and how we connect—so logically there is a burgeoning of information. Given the low costs of reproduction and distribution of information, there are *externalities* emanating from the creation of any single document. Individuals do not recognize the crowding effect of their contribution to the information pool. From the consumer side, a quest for the latest information and the fear of missing out creates *network effects*, enlarging the value of content when the network grows. Consequently, freed from the boundaries of time and space, and supercharged by externalities and network effects, connectivity and artificial intelligence have generated cognitive apathy and a sense of unease—an unimagined outcome. Where do we go from here and what are the guidelines? Who should set them?

At essence lies the illusion of free information. The Internet is the frontier analogous to the wild west. In order to tame the wild west of the Internet, the free model may have to be replaced. Social media sites unreservedly aggregate news from various curated media platforms, such as

© The Author(s) 2019
S. Bhatt, *The Attention Deficit*, https://doi.org/10.1007/978-3-030-21848-5_8

The New York Times, and reproduce them for free. The creator, researcher and writer of the story does not get paid for his efforts by these social media platforms, who are the publishers and distributors of content. The advertising business model is in charge, so perhaps we should begin with accurately pricing information, by using the subscription model. Users must pay a fixed fee for limited content. Google Search had originally instituted a "first click free" model, which allowed users to access the full text of the first article on Search. However, in June 2017, Google allowed individual subscription-based sites to independently determine their access policy.[1]

In addition to free access to information, we also have free creation of information—free online speech. Most people would agree that effective digital governance should include competitive markets supporting free online speech and respect for individual privacy. However, free speech and control over privacy create conflicting incentives. With 2.23 billion people using Facebook, there are roughly $(2.23b)^2$ possible connections. Which of these connections are active and how the network actually emerges is unpredictable. Connectivity, in this context, is supported by the 1st Amendment to the US Constitution, free speech. But then how does one define privacy in this free environment? If privacy is considered personal property, then it must be accorded property rights, which are 4th Amendment rights. If free speech intrudes upon personal space, then a potential dissonance arises between online enforcement of the 1st and 4th Amendments.

The design elements for global connectedness must recognize the malicious and perverse incentives and the unintended negative consequences created by a benevolent system design, according to Tim Berners-Lee on the 30th anniversary of his original proposal for an information management system (Berners-Lee 2019).

Malice in intent lies behind the recent surge in violent eruptions dotting the global landscape. On a domestic level, the digital ad campaign during the 2016 elections, which relied heavily on targeted advertisements

---

[1] In June 2017, Google ended the practice of granting "first click free" access to articles behind a paywall. The belief was that giving Google Search users access to articles would provide an incentive for subscriptions. *The Wall Street Journal* saw a drop in Search traffic after they ended this program, and Google Search now allows individual websites to determine how users access their sites from Search. https://9to5google.com/2017/09/12/google-search-first-click-free-ends/

based on analysis using Facebook data, misled guileless voters. Privacy missteps by Facebook with respect to the political data firm Cambridge Analytica's accessing users' data have led the US Federal Trade Commission to create a task force to examine potential anti-trust violations. Malice in intent covers online harassment, fraud and security breaches such as the 2013 breach of the retail giant, Target, which compromised the payment card information of up to 110 million customers. The root of the attack was malware installed on the point-of-sale systems so that credit and debit card information could be scraped as the cards were swiped (Rogers 2019).

Perverse incentives are created in the advertising business model. As discussed in Chap. 5, platforms, incentivized to increase viewer time spent on their site, support content triggering adversarial emotions over objective analysis and unbiased information, both of which can require greater cognitive bandwidth. This can lead to unintended consequences such as echo chambers, which reverberate with a sense of discomfort when facing unfamiliar views.

Unintended negative consequences of benevolent design include loss of privacy. Economic interdependence plays a role in the valuation of privacy. For example, explicit labor market contracts seem to rely on a "need-to-know" principle and only task-related information is provided. Implicit contracts, which arise in states of information asymmetry between employer and employee, may reveal additional information as signaling or screening mechanisms to more accurately assess and convey the employee's abilities. In contrast, a tightly embedded social network is likely to care less about privacy as sharing becomes the cornerstone of building self- and peer esteem. At the current scale of interconnectedness, control over personal information is rapidly being attenuated.

Laws and effective punishment can deter the first two problems, but the third problem, loss of privacy, can bedevil communities unless there is cooperation on a global scale.

There is a public good aspect to connectivity and information sharing on the Internet, as first discussed in Chap. 2. When the benefits of a good are felt by all, regardless of who is paying for them, the good is considered a public good. The benefits of shared health data are illustrated in the treatment of the Ebola virus epidemic in West Africa, when public health-care workers were apprised in a timely manner of the location and extent of the virus outbreak via Facebook-donated mobile satellite communication terminals, in addition to a donation notification on top of every Facebook user's News Feed. Shared traffic data allowed for efficiency in rerouting traffic, as was the case during the 2013 Pope Francis visit and the 2016 Olympics in Rio de Janeiro, using the Connected Citizens Program developed by Waze, a division of Google (Waze Launches Connected Citizens Program 2014). Shared data across agencies and governments could have helped prevent the 2019 attack in Sri Lanka. In this respect, the public good is epidemic containment, providing a safe environment for global public events and guarding against hate crimes. In fact, terrorism containment might be better addressed by using global connections to create a more just society and ameliorate the living standards of the perpetrators themselves.

"To get this right, we will need to come together as a global web community" writes Berners-Lee (2019). Robert Thomson, chief executive of News Corp, which owns *The Wall Street Journal*, feels that we are finally "discussing more seriously the fine lines between engagement and addiction, between repurposing and piracy, between belonging and bullying, between identity and insecurity, all of which are magnified digitally" (Robert Thompson 2019).

So we return to the drawing board and continue the conversation at a more abstract level. What *are* the potential benefits of connectivity? What exactly is the trade-off we are willing to make in exchanging privacy for more connectivity? Broadly speaking, we want the Internet to enhance the human experience defined by respect for individual dignity, basic human rights (which precludes malicious intent and perverse incentives) and free speech, and so we value privacy, transparency, inclusion, literacy, agency and innovation. We need privacy to preserve dignity, we need transparency to preclude malicious intent and perverse incentives and we need inclusion and literacy to ensure free speech. To innovate we need to preserve privacy of mind and allow for more "skin-in-the-game" or agency in risky choices so that creators of ideas reap the rewards, often non-pecuniary benefits.

Privacy

The weak link in the Internet system isn't a lack of government oversight, but our own gullibility (Berners-Lee 2019). E-commerce is convenient—time saving. Time is the important consideration when we divulge our private information in order to conduct business online or to check out salacious details about our friends. We want quick answers and are willing to part with personal information. Time is the driver when we relinquish agency and follow recommendation algorithms; we want our choices to be made for us in order to save time thinking about these decisions. A March 2019 survey found that 61 percent of US Internet users were either somewhat concerned or not too concerned or not at all concerned about privacy and security on Facebook. The equivalent number for Instagram was 76 percent (EMarketer 2019).

The relevant metric of connectivity is time. Our worldview is through the lens of time rather than through connections, both friendly and adversarial. The question of privacy is trivialized when convenience dominates and partial control over personal data is considered sufficient.

Privacy entails trust, responsibilities, reputation and rights. Weakly linked communities are likely to be low-trust societies, and so inbred norms promote privacy as a right. Revealing personal information would provide an adversarial neighbor with ammunition to settle scores or make tighter bargains. Reputation becomes a tool for harassment, making privacy even more valuable. Discrimination in actions and prejudice in beliefs are easier when you don't know your neighbors well. Responsibility for sustaining common resources and public goods diminishes since each individual prefers to free ride on the efforts of others. And then the notion of property rights and common civilian rights is threatened.

But privacy is not free. The price we pay for privacy will include limits on free content distribution as a shared resource in times of stress, balkanization of content and imperfect barriers to hateful content or safety systems. This is the second paradox of connectivity: a reversion to fragmentation and localization of ideas. The first paradox, discussed in Chap. 3, was solo consumption in a connected world.

Significant distributional effects are also likely. In order to enforce privacy, free consumer data can no longer support the advertising supported business model. Replacing free content with paid subscriptions to content platforms may create wealth or income-based ladders of knowledge, reinforcing the already massive inequality in income distribution. On the other hand, a self-selected use of consumer data to entice advertisers may provide just the resources to sustain free content. Perhaps personalization doesn't quite need long-lived metadata.

As a corollary, it would appear that high-privacy societies lack a motive for cooperation and sharing of information in the interest of social welfare and provision of public goods. However, the problems facing society today are of a global nature—climate change, terrorism—so cooperative mechanisms enlisting all humans as participants in the common solution are vital. The public good nature of information makes sharing personal data a common resource for handling pandemics, terrorism, epidemics and public health disasters, coordinating the logistics of traffic flows.

In high-trust societies, homophily plays a role in building trust and diminishing the need for privacy. Individuals are more likely to trust people with similar traits, belonging to the same peer group. Peer esteem is built by sharing traits with the group (see Chap. 3). There is less concern for privacy, as sharing is a way of legitimizing inclusion in the group. Concern for reputation correspondingly falls since individual information is mostly common knowledge, but there remains the deep responsibility to abide by the group norms. Inclusion into the group entails granting individuals rights over common property and public goods and, consequently, cooperation is a natural outcome.

Free online speech in the form of messaging and personal stories is the fastest growing aspect of connectivity and the private sector is addressing the ensuing privacy concerns. But caution against the theatrical amplification of antagonism and adversarial speech is needed. Seeking approval, the deranged mind follows a bizarre blueprint "in a lineage of hate crimes carried out for a captive audience of digital onlookers" (New York Times 2019).

The public mind is subject to associative heuristics turning a white supremacist incident in Pittsburgh, involving the killing of 11 *Jewish* worshippers at a synagogue in September 2018 into a touchstone of action against *Muslims* in New Zealand in March 2019, involving the massacre of 50 Muslim worshipers, and against *Catholics* on Easter Sunday in 2019, killing over 300 attendees at a church service in Sri Lanka. The Islamic State appears to be involved in the latter shootings, quite possibly in retaliation for the New Zealand attack.

Indiscriminate attacks on groups across the world present a trade-off between privacy and government surveillance. The safety and security of the disaffected and also disenfranchised requires "connecting the dots among different intelligence agencies and nations" (Washington Post 2019). The needs are larger than local politics; they are about providing necessities for a gainful human experience.

Social media can play a positive role in this effort. Yes, perpetrators of violence have used Facebook for disseminating their message, but these platforms can also be used to coordinate efforts across responsible institutions. Differentiating between public and private social networks, Facebook has recently added end-to-end encryption on two apps, Messenger and Instagram, to buttress the walls of private communication networks (WhatsApp, its messaging platform used extensively in South Asia, is already encrypted). Zuckerberg says "people sometimes want to interact in a town square, and sometimes they want to interact in the living room" (Nicholas Thompson 2019).

On the government side, the EU adopted the General Data Protection Regulation (GDPR) in April 2016, which was then implemented on May 25, 2018. GDPR is a European Union law on data protection and privacy for all persons within the Union and the European Economic Area. Firms that handle personal data have to recognize that these data are private property and obtain explicit consent from the data subjects prior to collection or usage.[2] These regulations might pose a larger burden

---

[2] The privacy principles underlying General Data Protection Regulation (GDPR) maintain that personal data must be

(i)   processed lawfully, fairly and in a transparent manner in relation to the data subject;

(ii) collected for specified, explicit and legitimate purposes and not further processed in a manner that is incompatible with those purposes;

(iii) adequate, relevant and limited to what is necessary in relation to the purposes for which they are processes

(iv) accurate and, where necessary, kept up to date;

(v)   kept in a form which permits identification of data subjects for no longer than is necessary for the purposes for which the personal data are processed; and

on startups and small businesses that don't have the financial resources to maintain compliance, compared with the behemoths such as Facebook and Google. While firms were made aware of these rules two years prior to implementation, there may still be a lack of understanding, leading to further administrative costs. These laws have served as a blueprint for laws elsewhere such as the California Consumer Privacy Act of 2018.

Transparency

Transparency is compromised by fake news, deceptive advertisements and phishing by bots. To be clear, the Internet has allowed the dissemination of misleading information, not invented it. There has always been the incentive to embellish stories in order to gain acceptance and stature, to increase audience size and to be remembered. ICT both increases the reach of content and shrinks the time to achieve this reach. So the time factor is critical. ICT viralizes all kinds of human activity but doesn't enhance human capacity to internalize or process activity. We may develop the technology to accelerate cognitive processing but have not done so *yet*. Hence, misleading information can have vast negative consequences due to the sheer scale of information transfer. The Cambridge Analytica scandal is not exactly one of misleading information but rather of manipulating users' behavior by using their personal data to provide them with exactly the kind of information that would elicit desired behavioral outcomes (Nicholas Thompson 2019). To be clear, data leakage from a psychology professor at Cambridge University, Aleksandr Kogan, to a developer, Cambridge Analytica, is the source of the problem, but actions taken by the final developer took center stage. John Rust, the head of Cambridge University's Psychometrics Center feels that

---

(vi) processed in a manner that ensures appropriate security of personal data, including protection against unauthorized or unlawful processing and against accidental loss, destruction or damage (Furman 2019).

artificial intelligence has a psychopathic personality.[3] "It's adept at manipulating emotions but underdeveloped morally" (Lapowsky 2018). So one could call this selective information provision.

The issue of transparency is paramount in any discussion of artificial intelligence technology. There is a growing literature on interpretable machine learning (ML) such that there is better understanding of complex machine learning models and data structures, leading to the construction of interpretable decision rules. For example, in the US a defendant is arraigned after arrest in a court appearance and provided with the charges made by the prosecutor. Subsequently, a judge decides whether to release the defendant on his own recognizance (RoR) or be subject to bail. In the latter case, many defendants remain in jail not having sufficient financial resources to post this bail. One of the most common pre-trial risk assessments is the Arnold Foundation's Public Safety Assessment (PSA). This algorithm scores a defendant's possible risk by evaluating nine factors.[4] A defendant with a prior violent conviction gets points. The PSA score can be used to estimate a defendant's likelihood of failing to appear for their court date or their risk of committing a violent crime. This one-size-fits-all approach is not appropriate for all jurisdictions since it ignores local idiosyncrasies. Access to data varies across cities, which measure risk factors in different ways and have variable preferences in their ranking of violent crimes.

AI facilitates the incorporation of a large number of variables in each jurisdiction obtained from thousands of pre-trial detention decisions, to assess the predictive power of different factors, in terms of flight risk or future crime. Machine learning can discover and synthesize patterns in the data and create a risk assessment tool that is safer and more equitable. However, machine learning systems are typically black boxes, which mean we don't understand how they work. In a criminal justice setting, a machine learning algorithm that factors residence as information could easily mistake

---

[3] Psychometrics is the science of assessing individuals' mental and psychological strengths and weaknesses. Applications of this science include personality tests, IQ and aptitude tests. A well-known paper by Kosinski et al. shows that easily available records from Facebook such as "likes" can be used to measure and predict opinions, personality, intelligence and sexual orientation (Kosinski et al. 2013).

[4] See the discussion in Chap. 2 on the role of bias in the criminal defense system.

that as a proxy for race in ways that reinforce bias. Transparency requires explanation of how the risk scores are obtained as a guard against such unintentional bias. Sentencing judges equipped with clever information can identify and release low-risk defendants and provide them with guidance for behavioral health assistance when deemed necessary.[5]

Inclusion and Literacy

Groups with more cohesion find it easier to participate in collective action, as peer esteem is built by identifying with groups and participating in collective action reinforces this sense of inclusiveness (Akerlof 2016). Group inclusion among the marginalized and disenfranchised is illustrated by the Easter 2019 terrorist attack in Sri Lanka. The incident appears to be the result of a partnership between the Islamic group ISIS and a small satellite group of Muslims. How else would such a small, ill-equipped group acquire the skills and weaponry to mastermind such a sophisticated attack? Previously, this group might have defamed statues. In the digital age well-funded global organizations can connect via the Internet and circumvent the well-secured borders of wealthy nations, attacking peripheral nations lacking the sophisticated anti-terrorist armor required in the current era.[6]

Inclusion and literacy are joint products to the extent that we learn best when we are included in the classroom. When some segments of the population are excluded from the world of the Internet via price discrimination, then there is no universal language, nor is there any learning. But is this any different from the pre-Internet world? A premier education is costly and accessible to the privileged few, but scholarships are a way of including the

---

[5] Jung et al. show that "simple rules are competitive with state-of-the-art machine learning algorithms." They explain their rules in the context of judicial decisions in pre-trial release. "Based on an analysis of over 100,000 cases, we estimate that judges can detain one-third fewer defendants, while simultaneously increasing the number that appear at their court dates." In the judicial context, as in many policy settings, it is statistically challenging to evaluate decision rules based solely on historical data. The key difficulty is that one cannot observe what would have happened under an alternative course of action (Jung et al. 2019).

[6] Fragmentation is enhanced when ostracized nations create their own internal network. The Halal Internet, started in 2011, resembles China's Great Firewall and is the Iranian government's controlled version of the Internet, called the National Information Network.

marginalized set of learners. The difference is that on the Internet, pricing can be opaque due to its multi-tiered nature, and therefore it is hard to recognize any form of discrimination. But it nevertheless makes accessibility difficult for those with limited ability to pay. For example, data pricing has at least three dimensions: speed, data rate and a cap on maximum data downloaded or streamed. For example, suppose a data plan for $100/month entails a bandwidth of 60 mbps and 2 GB cap on download or streaming. It also has an implicit data rate or throughput depending upon traffic. Hundred dollars might not withstand the budget of many low-income households if broadband access is considered a luxury good.

When differentiating products by price, or versioning as in the subscription model which charges the highest fees for a premium version, low-income groups may not be able to avail of all relevant information. In fact, since information is an experience good, there is knowledge of the worth of these higher-priced goods. For a budget-conscious household, it is simpler to consume the free, ad-supported content rather than the curated, but higher-priced ad-free content.

For example, Spotify is a major player in the streaming music market, not a music producer, and can play an important role in song and artist success. Decisions about which songs to promote are almost equivalent to editorial oversight at regular media sites and hence come with tremendous cultural power (Aguiar and Waldfogel 2018). Product differentiation is thrust upon users instead of allowing customers to independently choose from a collection of playlists. How does someone who is only familiar with Mozart ever learn about Migos?[7]

Agency

Agency is the salient factor in addressing cognitive apathy. While we may think that we are in control over *what* we peruse over the Internet, and indeed *when* we peruse, there is limited freedom in this regard. AI has become adept at pattern recognition by pooling large data sets and using

---

[7] The curated playlists on Spotify, Today's Top Hits and New Music Friday, introduce new tracks but simultaneously reward the artist. If an artist's track is placed on the curated playlist Today's Top Hits, the song has an increase of 20 million streams, while being ranked #1 on New Music Friday raises the song's streams by 14 million (Aguiar and Waldfogel 2018).

sophisticated algorithms to match individual attributes to outcomes, thereby segmenting users into ideological groups. Behavioral economists call this stereotyping, but effectively these algorithms are, for example, classifying individuals such as myself as a typical *The Economist* magazine reader and therefore feeding me information that an educated, East Coast liberal would appreciate. This is efficient, it saves me time. But it also shuts off whole chunks of the Internet world, areas that I cannot even imagine until some serendipitous encounter with a Breitbart follower alerts me to another way of thinking.

Social isolation, group think, loss of creativity and atrophy of human intellect are the consequences. According to a 2018 survey by the Edelman Group of 1000 US adults from the general population and 300 executives in technology roles, more than 70 percent agree that AI will lead to greater social isolation and less face-to-face communication.[8] Seventy-seven percent of the general population and 65 percent of tech executives also believe that devices with AI technology will lead to a *"dumbing down of people leading to a loss of human intellect and of creativity, with potential for digital group think"* and importantly, *"We are likely to become less inquisitive and more trusting of the information we are provided as accurate and authoritative [italics mine]."* Also, 74 percent of the general population, and 72 percent of tech executives, believe that AI will decrease the need for social interaction, leading to more isolation. Daniel Weld, a professor at the University of Wash writes, *"I worry that human abilities may atrophy [italics mine]"* (Edelman Survey 2019).

Innovation

Technology firms have grown large, becoming behemoths in content space. Could the rapacity of these behemoths be the actual cause of the dearth of small firms? In a low interest rate environment, large firms built on the twin pillars of network effects in demand and increasing

---

[8] The questions in the survey were developed by the Edelman AI Center of Expertise and augmented with input from the World Economic Forum. The general population was equally distributed from around the country and selected irrespective of their work positions, while selection of the technology executives focused on those who held senior management and C-Suite positions in their organizations.

returns to scale in supply are swallowing up incipient businesses valued around $100 million even before they can grow into mature firms valued over $1 billion (Liu et al. 2019). And as Tim Wu argues, "The FTC [US Federal Trade Commission] needs to get past its laser-focus on consumer prices, and figure out what competitive harm means in today's tech markets" (Wu 2018). The task force appointed by the British government and led by Jason Furman concludes that the behemoths' threat to competition must be dealt with a light hand such that we must give "every chance for competition to succeed in digital markets, tackling the factors that led to winner-takes-most outcomes and to that position becoming entrenched" (Furman 2019). The report recommends a new regulator, who would classify firms as gatekeepers due to their influence over a significant segment of the economy and would jointly develop a code of conduct. A large fraction of tech firms is headquartered outside Britain, but this report could serve as a blueprint for the US Federal Trade Commission and the European Commission.

References

Aguiar, L. and J. Waldfogel, 2018. *Platforms, promotions, and product discovery: evidence from Spotify playlists.* Cambridge: NBER Working Paper June 2018.

Akerlof, Robert. 2016. We Thinking and its Consequences. *American Economic Review: Papers and Proceedings* 2016, 106(5).

Berners-Lee, Sir Tim. 2019. *30 Years On, What's Next #ForTheWeb* Accessed 3/15/19 from https://webfoundation.org/2019/03/web-birthday-30/

Edelman Survey, 2019. Accessed 3/20/2019 from https://www.edelman.com/sites/g/files/aatuss191/files/2019-03/2019_Edelman_AI_Survey_Whitepaper.pdf

EMarketer, 5/29/2019. Accessed on 5/30/19 from https://www.emarketer.com/content/facebook-s-clear-history-tool-what-advertisers-should-and-should-not-worry-about?ecid=NL1001

Facebook joins the fight against Ebola with News Feed Donation Drive, The Verge, Nov. 6, 2014. Accessed 4/28/2019 from https://www.theverge.com/2014/11/6/7166725/facebook-joins-the-fight-against-ebola-with-news-feed-donation-drive

Furman, Jason. 2019. *Report of the Digital Competition Expert Panel: Unlocking Digital Competition*. March 2019. Accessed on 5/23/19 from www.gov.uk/government/publications

GDPR, Article 5, GDPR. Accessed 3/15/2019 from https://gdpr-info.eu/

Jung, Jongbin, Sharad Goel, Connor Concannon, Daniel Goldstein and Ravi Shroff. 2019. *Simple Rules for Complex Decisions*. Stanford University working paper. Accessed 3/22/19 from https://5harad.com/papers/simple-rules.pdf

Kosinski, Michal, David Stillwell and Thore Graepel, April 9, 2013. Private traits and attributes are predictable from digital records of human behavior. *PNAS*, 4/9/13. Accessed 4/24/19 from https://www.pnas.org/content/110/15/5802

Lapowsky, Issie. 6/19/2018. The Man Who Saw the Dangers of Cambridge Analytica Years Ago. *Wired Magazine*. Accessed 4/24/2019 from https://www.wired.com/story/the-man-who-saw-the-dangers-of-cambridge-analytica/

Liu, E., A. Mian and A. Sufi. 2019. *Low Interest rates, market power and productivity growth*. Cambridge: NBER Working Paper 25505. January.

NYT 4/28/2019. Accessed 4/28/19 from https://www.nytimes.com/2019/04/28/opinion/poway-synagogue-shooting-meme.html?utm_source=newsletter&utm_medium=email&utm_campaign=newsletter_axiosam&stream=top

Rogers, Phoebe. June 2019. *Breaching the Perimeter: An Exploration of Patters in Data Breach Environments 2005–2018*. Senior Thesis presented to the Department of Economics, Princeton University, 2019.

Thompson, Robert. 2019. Taming the Digital Wild West, *WSJ* editorial, 3/13/2019.

Thompson, Nicholas. 2019. 3/6/2019. Mark Zuckerberg on Facebook's Future and What Scares Him Most. *Wired Magazine* Accessed 4/24/19 from https://www.wired.com/story/mark-zuckerberg-facebook-interview-privacy-pivot/

Waze Launches Connected Citizens Program, Debuts Inaugural W10, PR Newswire, October 2, 2014. Accessed 4/28/19 from https://www.prnewswire.com/news-releases/waze-launches-connected-citizens-program-debuts-inaugural-w10-277867931.html

Washington Post. 2019. Accessed 4/28/19 from https://www.washingtonpost.com/opinions/global-opinions/sri-lankas-bombings-came-at-a-time-of-terrorism-fatigue-its-time-to-break-out-of-that/2019/04/23/0dda7890-660e-11e9-a1b6-b29b90efa879_story.html?utm_term=.89ce1ee51522&wpisrc=nl_opinions&wpmm=1

Wu, Tim. 2018. *The Curse of Bigness: Anti-Trust in the New Gilded Age*. Columbia Global Reports.

9

Conclusion: Dialogue, Not Walls

Summary

The digital revolution ushered an era of connectivity and then led to fear, mistrust and risk avoidance. What happened and what do we do about it? Dramatic changes in the environment require adaptation and restructuring of one's mental frame of reference. A sudden change in the weather during an outdoor wedding event, when tables are laid with pink florals and green linens, requires rapid rearrangement of plans in addition to eliciting the cooperation of the assembled guests. This is effortful, utilizing cognitive bandwidth. It would require less mental effort to plan an indoor wedding than to exercise such concentrated recalibration. Risk avoidance is the outcome because cognitive effort required to navigate changing circumstances exceeds the pleasure from a beautiful outdoor setting.

A tsunami of content, triggered by the sharing model, overwhelms our attention to resources so forcefully that cognitive apathy provides a refuge. Attention deficit thwarts the capacity to think, so we resist change and avoid risk. Nostalgia, for instance, is a natural reaction to frequent change and an unknown, uncertain future. By identifying with a known past, a tranquil environment where change was slow, individuals feel a sense of belonging and are better able to regulate their current behavior. Nationalist sentimentality is an example of this phenomenon.[1]

ICT orchestrated both connectivity and asynchrony. Connectivity across humans and machines unleashed the sharing model. The decoupling of time and content allows asynchronous communication and therefore solo consumption. Sharing has two faces: the trusting side where transparency and openness are dominant and the fearful anxious side where judgment, comparison and mistrust become apparent. The Internet was created with the trusting, risk-seeking face, but it has subsequently morphed into narcissism and mistrust, led by a focus on the self, and the *in-group*. The paradox of connectivity is wall-building, privacy enhancements and prejudice amidst a globe traversed by communication channels. Decision-making now involves choices that require little adaptation to changing circumstances, which means that there is an overall resistance to change. Cognitive apathy makes people view themselves as less open to changing outcomes, less entrepreneurial and therefore less risk-taking.

Free Will Versus Cognitive Apathy

Human civilization has never encountered such massive connectivity—over half of the world is connected: 56 percent of the global population consists of Internet users and 88 percent of the US population is thus connected. Almost a third of the world is on Facebook. But the evidence laid out in Chap. 6 creates a different worldview: low labor productivity, a startup deficit, an increase in mortality rates due to suicide and drug overdose and an increase in mental stress across the post-boomer demographic. Gordon (2016), citing low labor productivity statistics, is cautious about the economic impact of ICT, while concern over devices themselves and the digital method of transmission has

---

[1] A simple syllogism captures the logic:

Content tsunami $\longrightarrow$ attention deficit and cognitive apathy

Cognitive apathy $\longrightarrow$ less cognitive bandwidth available for adaptation to change, so there is resistance to change, which manifests as risk avoidance $\longrightarrow$ less risk-taking behavior

Therefore, content tsunami $\longrightarrow$ less risk-taking behavior

been repeatedly articulated. There is some evidence suggesting that the device providing content—smartphones—is itself suspect. Experiments on smartphone usage have shown that "the mere presence of these devices reduces available cognitive capacity" (Ward et al. 2017).[2]

With connectivity comes sharing of thoughts and ideas, a massive exchange of data. The ensuing tsunami of content stimulates comparisons and judgments clogging mental data-processing capabilities and demoralizing individual sensibilities. Cognitive apathy arises as self-protection against mental stress and overload. A pervasive "do-nothing" attitude is reflected in diminished risk-taking and lack of bold action. We observe this in business and everyday life.

Connectivity across both man and machine has created two mutually conflicting forces: sharing and comparing. Transparency, trust and cooperation are the waves of the sharing stream, while judging, comparing and retreating behind self-constructed walls of private gardens are the dams across this stream. Bold risk-taking follows from trust that the world will eventually figure things out. Walled gardens, on the other hand, stoke narcissism, apprehension, fear and risk avoidance. The world is appraised with a sense of dread and anxiety in an in-group versus out-group vision, and cognitive apathy provides a comfort zone—a space in time.

Harare writes "since the global financial crisis of 2008 people all over the world have become increasingly disillusioned with the liberal story. Walls and firewalls are back in vogue. Resistance to immigration and trade agreements is mounting." And this disillusionment has created a sense of cognitive dissonance with connectivity itself. We can connect but is this good for us? Harare elaborates that liberal elites have "entered a state of shock and disorientation" (Harare 2018; Carr 2008).

When individual liberty is built on a foundation of personal dignity, consisting of self-esteem and agency, then freedom in a connected world can foster narcissism and tribal *groupy* thinking with comparisons and judgment. Fear of the "other" and losing one's status tend to elicit risk avoidance and discrimination emanating from a need for privacy coupled

---

[2] Carr feels that consumption of digital content is at a speed that precludes deep processing (Carr 2008). This argument relies on understanding the complexities in brain circuitry, and hence it is ambiguous until we have more research in neuroscience.

with increasing prejudice. Balancing this freedom in a civilized society is the sharing model, which defines the human experience in terms of stories and imagined realities so as to generate a sense of belonging for all—a leveling of the playing field. *The American Dream.* Sharing pictures and text, building connections, to create a narrative or imagined reality is what *keeps us going*, what motivates us. The balancing act between freedom, privacy and distrust on the one hand and a shared vision of cooperative storytelling on the other is delicate.[3] How can these stories, this content, overturn a shared vision? I posit that the content tsunami has made this vision inchoate. Having lost our way, mistaking the trees for the forest, we are overwhelmed with cognitive apathy to not recognize it.

The search for meaning in human existence led us to notions of identity and legitimacy, of being acknowledged. Operationally, that means aligning with groups to create a sense of community, of peer esteem. There is a need for social networks, for community, to recapture agency in order to foster a sense of belonging and stability. It is possible that "the neglected third pillar, the community," has been forsaken by multilateral institutions convening on the global stage, interceding on behalf of their respective populations (Rajan 2019).[4] The concierge mode of operation where decision-making is delegated to parties with no skin-in-the-game creates lost opportunities as granular information pertinent to that society is lost. State authorities, for example, could not make appropriate educational placement of my autistic son when he was in second grade—they needed guidance from his kindergarten teacher and his parents.

How do communities make decisions wisely when individuals are overwhelmed with cognitive apathy? Engendering self-determinism requires fulfilling individuals' need for belonging but how does one curb the sharing model already unleashed on the global stage? Moreover, the

---

[3] There is a vast literature debating the compatibility of individual freedom and democratic notions of equality of opportunity. Democracy, for some, was the idea that people should be free to make the most of their abilities in an egalitarian society (Alexis de Tocqueville, Rawls, Jeremy Bentham, John Stuart Mill), which implies that we should provide the resource (via taxes) to enable this outcome. But then there is the opposing idea that people own their talents and cannot be compelled to share its rewards, that is, via taxes (Robert Nozick).

[4] In the model of former International Monetary Fund (IMF) chief economist Raghu Rajan, the first two pillars are the state and markets (Rajan 2019).

power of AI and biotech may erode individual rationality or free will and reinforce cognitive apathy. Recall the discussion of preferences formation in Chap. 1, where we questioned the notion of consumer sovereignty and recognized that personality and tastes can be influenced by sharing information. If emotions and feelings guide decision-making, then AI together with biotech can reverse engineer the mind. Evolutionary psychologists have found that feelings are biochemical processes, evolved over millions of years of evolution. Neurons and synapses touched by sensory inputs and processed by biochemical reactions generate feelings. Harare believes that one could therefore manipulate these inputs to generate the requisite emotion and the desired decision. A liberal, free market democracy "will mutate into an emotional puppet show" (Harare 2018).

> If by "free will" you mean the freedom to do what you desire—then yes, humans have free will. But if by "free will" you mean the freedom to choose what to desire—then no, humans have no free will. (Harare 2018)[5]

Manipulation of our feelings is not new; access to enormous data which serve as inputs into the feelings machine is the novel part. Since individuals are incapable of flawless data processing and accessing complete data, they cannot know their own feelings and therefore they cannot know themselves. But humans have become *data cows*, or "tame humans that produce enormous amounts of data and function as very efficient chips in a huge data-processing mechanism" (Harare 2018).

The significant question about free will becomes one about thinking. What is the nature of thought processes? Is it computation, data analysis and storage? Or is there something more? What is the essence of human cognition? Is there proof that there is more? The word *ambiguity* surfaces often in discussions about the uniqueness of the human intelligence compared with machine intelligence. Prediction using data can then replace human cognition and, therefore, free will.

---

[5] "Many people, including many scientists, tend to confuse the mind with the brain, but they are really very different things. The brain is a material network of neurons, synapses and biochemical. The mind is a flow of subjective experiences, such as pain, pleasure, anger and love" (Harare 2018).

But then who owns the data? The business model of the attention merchants, where people are the data, organizes this vast quantity of data by creating a classification system. Data are sorted and classified into intelligible categories, created to make these categories manageable without infusing them with deeper meaning. In so doing much diversity and cultural context is neutralized, as when Spotify's musical genres classify music into either hip-hop or pop music forcing artists to repackage their art into self-defined categories to help fans decipher how to listen to music. Amazon's product categories and individual profiles on various media sites lump consumption and people in a manner that creates new context, new identities for people, obscuring subtle differences and fostering stereotyping. However, the gain in preserving context must be measured against the loss of algorithmic ease. A picture of the economy created by broad statistics such as GDP, unemployment and so on. GDP does not capture actual consumer happiness nor do wages capture worker well-being on the job, but it helps in managing macroeconomic policy.

By defining precise categories, we might be facilitating a robotic interpretation of our race, whereas an element of vagueness purportedly supports our humanity. The printing press altered human interactions by replacing blind faith with reason and individual analysis. And now, data and AI, together, can decipher and manipulate individual preferences and personalize recommendations. Consequently, we develop unquestioning trust in devices and software. Embedded in a world of incessant chatter even while we consume in solo mode, attention scarcity leaves us little time for reflection. Then the very ambiguity that humanizes us is lost.

As humans are sliced and diced as data points, sorted into categories based on homophily and guided in decisions by machines, there arises a fear of drowning in a sea of conformity and oneness. We grapple with the question of identity (recall the discussion of identity in Chap. 3). In truth, real freedom or personal dignity is having the space to align with any group and carve out one's difference from others to build peer and self-esteem. Free will is having the agency to choose one's social category and behave accordingly (Akerlof and Kranton 2008). Attention scarcity confounds this choice, so we observe cognitive apathy as a defense against the ensuing uncertainty and ambiguity.

Bold Ideas and Cooperation

In ancient Athens, Aristotle's hangout, city-states were tightly knit communities and public versus private distinction did not exist. All citizens shared the same goals and values, so the pursuit of happiness is a cooperative enterprise. Citizens cooperate not because the law compels them to, but out of a friendship that comes from sharing their lives and goals.[6] Robert C. Solomon summarizes the "Aristotelian view that we are first of all members of a community and our self-interest is for the most part identical to the larger interests of the group. Competition [in business] presumes, it does not replace, an underlying assumption of mutual interest and cooperation" (Solomon 2004). Central to the notion of cooperation is trust. "Creating trust is taking a risk. Trust entails lack of control, in that some power is transferred or given up to the person who is trusted." The notion of the individual as separate from the community was invented in twelfth-century Europe after periodic wars broke families and feudal structures. Self-interest and lack of cooperation reduces trust, makes people more suspicious and risk averse.

Human power rests in mass cooperation built on shared beliefs or an imagined reality. There was a time when religion provided just such a cooperative mechanism, when millions of people believed in a common story for thousands of years. In practice, the power of human cooperation depends on a delicate balance between truth and fiction (Harare 2018). Like money, people began to sanctify the Bible or the Vedas only after long and repeated exposure.

> Humans rarely think for themselves. Rather, we think in groups. What gave Homo Sapiens an edge over all other animals and turned us into the masters of the planet was not our individual rationality but our unparalleled ability to think together in large groups. (Harare 2018)

---

[6] "According to Aristotle, one has to think of oneself as a member of the larger community—the Polis for him, the corporation, the neighborhood, the city or the country (and the world) for us—and strive to excel, to bring out what is best in ourselves and our shared enterprise" (Solomon 2004).

For millions of years, humans have been living in small bands of no more than a few dozen people. "Unfortunately, over the past two centuries, intimate communities have indeed been disintegrating. ... Consequently, people live ever more lonely lives in an ever more connected planet" (Harare 2018). Furthermore, the paradox of sharing experiences is that the rewards become extrinsic rather than intrinsic to personal satisfaction and meaning. Traditional gathering places, such as shopping malls, are disappearing, reinforcing the use of screen time as the mode of choice for socialization. Asynchronous connectivity actually increases social isolation with corresponding effects for mental health, drug and alcohol addiction, and more generally in cognitive apathy, as documented earlier in Chap. 7.

In the Savannah over four billion years ago, our ancestors learnt to share information and cooperate, perhaps to overcome the futility of competing against the forces of nature, in order to survive. In the modern era, the human mind has evolved to create notions of private property. Associated with property rights is competition over resources, not for survival but for status. While global connectivity has produced sharing, this is not cooperation. Cooperation involves recognizing our mutual dependence and our shared problems. "*We cooperate in order to better comprehend*" our own problems and so we share (Bhatt 2017).

Sharing must be coupled with judgment and appraisal of the implications. With cooperation, we are engaged together, but sharing must be measured. Access to all information is neither productive nor reassuring for people or organizations. Collective action works not because we subsume our preferences for the greater good, but because our preferences incorporate the larger social network in recognition of connectivity. And interdependence.

The twentieth-century vision of free markets in goods and people is scaled up in the twenty-first century to free markets in ideas due to constant and ubiquitous connectivity. Is the liberal vision of free markets in ideas, people and goods a sustainable model? This requires more than simply a democratic government, free movement of people (immigration) and a free press. We need coordination and cooperation toward a common vision. Scholars have promoted this idea centuries ago, as in the Aristotelian model of a social contract entailing cooperation, an obligation toward a common purpose. What is required is boldness in moral leadership, boldness in addressing common problems, strengthening the fragile threads that bind us as we enter this uncharted territory.

In the face of ecological disruption and massive climate change, perhaps cooperation is once again the best response. We can draw out the implications of connectivity, AI and biotech a few steps ahead, but engineering of both the outside and the inside worlds can lead to a sequence of consequences beyond our comprehension.

Behavioral social science and evolutionary psychology have made the case that human judgment and decision-making are biologic adaptations rather than engines of pure rationality. Therefore, *unchecked* information is *not* the perfect tool for addressing our problems of social justice, fair distribution of rewards and fair opportunity. We can create balance either by judiciously disseminating information or by randomly building walls to contain the transmission of information. Nation building in the modern era appears to have followed the latter path.

I believe that prudent information exchange can be a useful tool in developing the criteria for social justice by balancing notions of equality and freedom so that fair rules of the game are applied to socio-economic interactions in a free market. We, as a society, need to find a way to address the consequences of cognitive apathy—the decline in creativity, entrepreneurship and imagination.

Consider the power of information in the Native American West. Water from aquifers in the California's Coachella Valley is powering golf courses, glittering pools and a music festival in a region where temperatures hover around 120 °F. This desert land is the native home to the Agua Caliente Band of Cahuilla Indians who regained rights to this water in a 2017 Supreme Court decision reaffirming the lower courts' ruling in the tribe's favor. Approximately 50 percent of homes on Native American land have inadequate access to drinking water compared to 1 percent for all US homes.

Spreading information about this decision has prompted unresolved claims to move forward, such as the Navajo Indian reservation's claims to groundwater in northeastern Arizona and across the border in northwestern New Mexico (areas surrounding the beautiful destinations of Mesa Verde National Park) and northeastern Utah (Canyonlands National

Park and Arches National Park). Dissemination of this content is vital and could set a powerful precedent for tribal groundwater claims in many other states (Womble et al. 2018).

Hence, content can be constructive when used discriminatingly, recognizing that the salient information at distinct times and geographies can be different. The criteria for judgment are based on underlying ideologies such that thoughtful, judicious dialogue becomes vital in bridging different frames of reference in order to arrive at a consensual set of criteria for the judgment in question.

But first, we need to have a common language and a two-way interchange. ICT is a unique tool for disseminating relevant information and facilitating discussion. We can then have a dialogue about our differences and their resolution. Through shared thinking we can address the twin problems of climate change and resource allocation for global improvement and amelioration of human lives. Awareness of consequences might be the best way to dilute adversarial incentives which promote physical and digital misappropriation of resources.

Digital connectivity can inspire collective dialogue, consensual and constructive disagreement and a sense of mutual awareness. Dialogue is a two-way interchange, based on listening and appraisal. It is a bold vision of cooperation. In this audacity lies a risk appetite for taking chances, for pooling our thoughts to confront uncertainty. Sincere discourse could be a fortuitous consequence of digital connectivity.

References

Akerlof, George A. and Rachel E. Kranton. 2008. Identity, Supervision and Work Groups, *American Economic Review: Papers & Proceedings 2008*, 98 (2), 212–217 http://www.aeaweb.org/articles.php?doi=10.1257/aer.98.2.212

Bhatt, Swati. 2017. *How Digital Communication Technology Shapes Markets*. New York: Palgrave Macmillan.

Carr, Nicholas. 2008. Is Google Making Us Stupid? *The Atlantic*, July/August 2008. Accessed 3/22/2017 from https://www.theatlantic.com/magazine/archive/2008/07/is-google-making-us-stupid/306868/

Carr, Nicholas. *How Smartphones hijack our minds*. Accessed 3/22/2017 from https://www.wsj.com/articles/how-smartphones-hijack-our-minds-1507307811

Gordon, Robert J., 2016. *The Rise and Fall of American Growth: The U.S. Standard of Living since the Civil War*. Princeton: Princeton University Press.

Harare, Yuval. 2018. *21 Lessons for the 21st Century*. London: Penguin Random House.

Rajan, Raghu, 2019. *The Third Pillar: How the Markets and State Leave the Community Behind*. Penguin Press: New York.

Solomon, Robert C. 2004. Aristotle, Ethics and Business Organizations. *Organization Studies* 25 (6): Sage Publications. Accessed 4/29/2019 from https://journals.sagepub.com/doi/pdf/10.1177/0170840604042409

Ward, Adrian, Kristen Duke, Ayelet Gneezy and Maarten Bos. 2017. Brain Drain: The mere presence of one's own smartphone reduces available cognitive capacity. *Journal of the Association for Consumer Research*. 4/3/2017. Accessed 3/22/2017 from https://www.journals.uchicago.edu/doi/full/10.1086/691462

Womble, Philip, Debra Perrone, Scott Jasechko, Rebecca Nelson, Leon Szeptycki, Robert Anderson, Steve Gorelick. 2018. *Indigenous communities, groundwater opportunities*. Science 03 Aug 2018: 361 (6401), pp. 453–455. DOI: https://doi.org/10.1126/science.aat6041 Accessed 5/17/19 from https://science.sciencemag.org/content/361/6401/453.full

Index[1]

---

[1] Note: Page numbers followed by 'n' refer to notes.

© The Author(s) 2019
S. Bhatt, *The Attention Deficit*, https://doi.org/10.1007/978-3-030-21848-5